Columnar Structures of Spheres

Columnar Structures of Spheres

Fundamentals and Applications

Jens Winkelmann

Ho-Kei Chan

JENNY STANFORD
PUBLISHING

Published by

Jenny Stanford Publishing Pte. Ltd.
101 Thomson Road
#06-01, United Square
Singapore 307591

Email: editorial@jennystanford.com
Web: www.jennystanford.com

British Library Cataloguing-in-Publication Data
A catalogue record for this book is available from the British Library.

Columnar Structures of Spheres: Fundamentals and Applications

ISBN 978-981-4669-48-1 (Hardcover)
ISBN 978-0-429-09211-4 (eBook)

This book is dedicated to Prof. Denis Weaire, Prof. Stefan Hutzler, and Dr. Adil Mughal, who introduced us to the fascinating field of packing problems during our wonderful times at Trinity College Dublin.

And Happy 80th *Birthday to Denis!*

Contents

Preface

Columnar structures, many of which are helical, refer to dense cylindrical packings of particles. They are ubiquitous, for example they exist in the contexts of botany, foams, and nanoscience. There have been in-depth investigations of columnar structures of both hard spheres (e.g. ball bearings) and soft spheres (e.g. wet foams) through computer simulations, analytic derivations, or simple experiments. This monograph serves as a comprehensive guide for any scientist, engineer, or artist who would like to have a good grasp of the fundamentals and applications of such aesthetically appealing structures for his or her own professional interests.

This monograph is organized as follows: We first give an introduction to the field of packing problems, where such problems are not only related to the columnar structures presented in this monograph but also to the structures of condensed matter systems in general. We then discuss what columnar structures of spheres are, with an overview of their classifications and possible applications. This is followed by a discussion of the models and concepts employed in the study of such columnar structures. Following this, we discuss in detail a method of sequential deposition for generating columnar structures of hard spheres computationally or experimentally. We then present findings on columnar structures of soft spheres and on buckled columnar structures of longitudinally compressed hard-sphere chains.

This monograph is a collection of original research carried out by the two of us in the Foams and Complex Systems Research Group of Trinity College Dublin at respectively different eras [Ho-Kei Chan, post-doctoral research fellow (2009–2012); Jens Winkelmann, PhD student (2015–2020)] under the supervision of Prof. Denis Weaire and Prof. Stefan Hutzler and in collaboration with Dr. Adil Mughal from Aberystwyth University. We are grateful to Denis, Stefan, and Adil for introducing us to the fascinating field of packing problems during our wonderful times at Trinity College Dublin. We end this preface by sharing a few memorable pictures taken during our times at Trinity. May the Foams and Complex Systems Research Group of Trinity College Dublin continue to thrive for many years to come.

Jens Winkelmann
Ho-Kei Chan
Autumn 2022

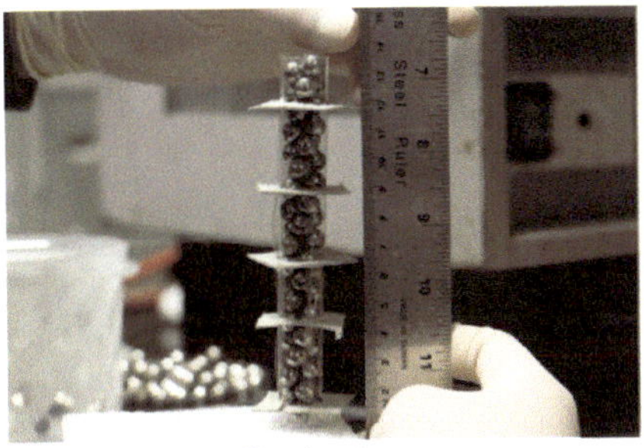

Experiment on
columnar structures of ball bearings
(Ho-Kei Chan, 15th October 2009)

Experiment on
columnar structures of foams
(Ho-Kei Chan, 27th June 2011)

Experiment on
columnar structures of sweets
(Jens Winkelmann, 26th May 2016)

Presentation on Rotational columnar structures
at Yale University
(Jens Winkelmann, 4th June 2019)

Stefan Hutzler
Aaron J. Meagher
Cathal B. O'Donovan

Matthias E. Möbius
Stephen J. Hardiman
Steven T. Tobin

Zigzag packing of colleagues
(Group photo in Belfast, 6th March 2010)

Acknowledgements by Jens Winkelmann

Since this monograph is in large parts based on my PhD research [Winkelmann (2020)] at Trinity College Dublin (Ireland), I'd like to start this acknowledgement by thanking everybody who was involved in making my post-graduate time all the more enjoyable and wonderful.

I had the great opportunity to take part at over 10 conferences during the last four years. It allowed me to travel to amazing places such as Italy, Cambridge, and Yale where I had the opportunity not only to meet old friends, but also to make new ones.

As mentioned in the preface, the three persons most involved in my PhD were Stefan Hutzler, Denis Weaire, and Adil Mughal. I'd liked to say thank you to Stefan Hutzler for his support as my PhD supervisor and Denis Weaire for all his great ideas which shaped many chapters of this monograph tremendously. Special thanks also go to Adil Mughal from Aberystwyth University, Wales, for a fruitful collaboration on studying columnar structures. He also proofread (and tore apart) parts of my thesis and thus also immensely impacted this monograph.

My thesis was also heavily improved by the critics of my PhD examiners, Gerd Schröder-Turk and Graham Cross. Their comments therefore also enhanced this monograph tremendously. I'd like to thank them for their thorough comments and criticism.

Big thanks also to my former office mates, Ben and Fritz! It was always great craic with you guys in and outside the office. In this regards, I also want to thank you two for helping to proofread two chapters of my thesis. These two chapters have become Chapters 5 and 6 of this monograph.

I would also like to thank Steven Burke and John Ryan-Purcell for their proofreading of two other chapters of my thesis. These two chapters have become Chapters 7 and 8 of this monograph.

Group photo after my PhD viva (4th November 2019)

Of course, I thank my parents and sister for all the support throughout all the years of studies.

A *huge* thank you also goes to Bernadette McAweeney for the great hospitality during the four years of my PhD, even though I only intended to stay for a few weeks, initially. I got treated like her fourth son at her house (not always to my own fortune).

Furthermore, I'd like to thank some old friends from home in Essen (Germany). `Tobey0r` (Tobias Jung) made my life as a PhD student easier in regards to design, graphics, and proof-listening many presentations before conferences.

I'd like to thank Tracy as well for her many visits and travelling with me all around Ireland. It helped me to stay sane in some stressful times.

I also ought to mention the helpful support by Maik Malki, Christian-Roman Gerhorst, and Ismo Toijala. Especially Ismo Toijala's amazing job: He proofread my complete thesis (and therefore huge parts of this book) within six hours and found 312 (minor) mistakes in the (almost) final version of my thesis.

Thanks also to my personal enemy no. 2, Marie Schmitz, for being mean and always failing to kill me. I hope you will suffer the same cruel fate as personal enemy no. 1!!!

Jens Winkelmann

Chapter 1

An Introduction to Packing Problems[1]

A stack of oranges on display at a grocery store represents a solution to a centuries-old problem: How can we pack identical spheres as densely as possible in an open space? An endless variety of such packing problems exists, including packings of spheres, spheroids, rods, deformable bubbles, and many other shapes, which may be packed in two, three, or more dimensions. Powerful computer techniques have opened up new avenues for solving packing problems. The results have helped to verify or disprove some very old conjectures, and they relate to the structures of matter and the efficient, secure encoding of information.

1.1 Packing Problems in Daily Life

Whoever you are, you probably have experienced some challenges of "packing" in your daily life. For example, when preparing for a flight, you might scratch your head wondering how to pack all of your belongings into your suitcase and carry-on bag without exceeding the weight limit for either bag. There is also a limit for the size

[1]This chapter is an updated version of a 2012 article by Ho-Kei Chan [Chan (2012)].

Columnar Structures of Spheres: Fundamentals and Applications
Jens Winkelmann and Ho-Kei Chan

ISBN 978-981-4669-48-1 (Hardcover), 978-0-429-09211-4 (eBook)
www.jennystanford.com

dimensions of your carry-on bag. So, if you are taking a relatively long object, you might need to pack it diagonally. Yet, such a move might affect how you pack your remaining items.

Packing problems, however, do not always cause troubles, and they can also be great fun. Just take a look at those sudoku addicts on a train or in a cafe who immerse themselves in the game trying to "pack" numbers into a 9×9 grid (with some pre-filled sites) in order to have all the numbers from 1 to 9 appear in every column, every row, and every 3×3 sub-grid. What a challenge! But like many other games, it is the pleasure of overcoming a challenge that causes the sudoku players to drown in the game.

Some other packing problems are at least as interesting as sudoku and, more importantly, they hold the key to an understanding of what makes up our physical world. Consider a basket of oranges—a grocer could show you how to pack the oranges into their densest possible arrangements on a table, and a physicist will tell you that the same arrangements are found for atoms and molecules in a surprisingly wide range of everyday materials.

1.2 Packing Problems in Physics

It seems that the scope of packing problems is wide and reaches almost all walks of life. But what exactly do we mean by packing problems?

In physics, a packing problem consists of finding optimal spatial arrangements of objects under some well-defined constraints, such as the shapes and sizes of the packed objects and those of the confining container (if there is one). Also, the meaning of the word "optimal" needs to be defined. For example, it might be specified as the maximum density or the minimum surface area. In the orange-packing problem, the oranges are (to a good approximation) spherical and equal-sized, they are to be packed in an open space rather than a confined one (such as a crate), and "optimal" means densest. In the case of sudoku, the numbers have to be packed into a 9×9 grid that already has certain numbers fixed in place, and "optimal" refers to the requirements about the spatial arrangement of all the numbers.

Figure 1.1 When oranges are stacked in their densest arrangement, each layer has two sets of interstitial sites (red and green), either of which can accommodate the next layer of oranges.

Let us return to the oranges. What are the densest possible arrangements of oranges in three-dimensional space?

A common way of stacking oranges is to make a layer of oranges arranged hexagonally (with every orange touching six other oranges) and then stack further hexagonal layers on top of it. The oranges on the second layer are placed over interstitial sites of the first layer. There are two sets of these interstitial sites, each forming a hexagonal pattern where the upper layer of oranges can be placed (see Figure 1.1). Depending on which set of interstitial sites you choose in each deposition step, you can obtain structures with different symmetries. For instance, the structure that repeats for every pair of adjacent hexagonal layers (ABAB...) is known as the hexagonal close-packed (HCP) structure, whereas the one that repeats for every trio of adjacent layers (ABCABC...) is described as face-centered cubic (FCC).

All these structures share the same packing fraction (the volume of the packed spheres divided by the total volume), namely $\pi/\sqrt{18} \approx 0.74048$, which was conjectured by Johannes Kepler in 1611 to be the highest possible packing fraction for identical spheres in three dimensions. We "know" from experience that this conjecture is true. However, the wait for a widely accepted mathematical proof lasted nearly four centuries, until when Thomas Hales used a novel computer-aided approach to exhaust the different possible configurations [Hales (2005)]. Quite a contrast to the time scale for solving a sudoku problem!

A similar scenario existed for Lord Kelvin, who conjectured in 1887 about the optimum packing of equal-sized (i.e., equal volume) soap bubbles in three dimensions, which is now known as the Kelvin problem [Weaire (1996)]. He conjectured that the structure with the least surface area was one that is made up of a particular type of tetradecahedron (i.e., a polyhedron with 14 faces) with six square and eight hexagonal faces. A century later, however, his conjecture was disproven by Denis Weaire and Robert Phelan, two physicists from Trinity College Dublin, who discovered a better structure [Weaire and Phelan (1994)] through computer simulations using the software Surface Evolver [Brakke (1992)]. The Weaire–Phelan structure consists of two types of bubbles that have the same volume but different shapes. For a given bubble volume, the structure's surface area per bubble is 0.3% less than that of Lord Kelvin's tetradecahedra.

Is the Weaire–Phelan structure the ultimate solution? We believe so. But wait—as was the case for Lord Kelvin's conjecture, a formal proof is still lacking and therefore we cannot yet eliminate the possibility of finding an even better structure.

1.3 Computational Aproaches to Packing Problems

Those two examples, of oranges and soap bubbles, demonstrate how difficult and challenging it can be to completely solve a packing problem. Thanks to advances in computer technology over the last few decades, many packing problems have been dealt with via a computational approach. In many cases, the solutions have not been rigorously proven in a strict mathematical sense, but with the development of various numerical tools a wide range of dense structures—many believed to be the densest possible—have been generated via computer simulations. The basic idea is to simulate a physical process of energy minimisation, with one or more types of interaction energy defined for the entities involved—somewhat like a network of connected springs seeking a configuration with the least stretching. The assumption is that the optimal structure, often referred to as the densest one, is one with the lowest total energy.

For example, in molecular dynamics simulations, an interaction energy (e.g., the Lennard–Jones potential) is defined for every pair of particles as a function of their separation. Typically, this interaction energy has a minimum at some equilibrium separation of the pair. A loose ensemble of particles will interact and then self-assemble into denser, spatially ordered clusters, because the interactive forces tend to drive every pair of particles toward their equilibrium separation.

Another well-known approach is simulated annealing, in which the process of energy minimisation involves only random displacements of particles but no equation of motion (unlike molecular dynamics simulations). In each computational step, the new configuration of the system, as obtained after some random movements of particles, is either accepted or rejected. A lower-energy configuration is always accepted, but a higher-energy configuration might still be accepted with a certain probability. The latter prevents the system from being trapped in a metastable state (a local energy minimum that is not lowest overall) and gives it a finite probability of arriving at its lowest-energy configuration.

These approaches, their many variations, and some alternative computational methods (e.g., sequential deposition [Chan (2011, 2013); Chan *et al.* (2019)]) have been employed to study the optimal packings for a rich variety of particles in bulk or in confinement, ranging from circles [Füredi (1991); Lubachevsky and Graham (2003)], spheres [Pickett *et al.* (2000); Koga and Tanaka (2006); Duran-Olivencia and Gordillo (2009); Chan (2011); Mughal *et al.* (2011); Srebnik and Douglas (2011); Heitkam *et al.* (2012); Mughal *et al.* (2012); Chan (2013); Mughal (2013); Wood *et al.* (2013); Mughal and Weaire (2014); Yamchi and Bowles (2015); Fu *et al.* (2016); Fu (2017); Fu *et al.* (2017); Mughal *et al.* (2018); Chan *et al.* (2019)], ellipses [Jin *et al.* (2021)], ellipsoids [Donev *et al.* (2004a); Odriozola (2012); Bautista-Carbajal *et al.* (2013); Jin *et al.* (2020)], to polyhedra [Torquato and Jiao (2009)]. In Section 3.5, we review some of these simulation methods that have been used in the context of columnar structures.

1.4 Random Packings of Particles

Many structures obtained from these numerical schemes are crystalline (meaning they follow a regular, periodic pattern), or at least display long-range order (meaning that some regularities persist over long distances). But also interesting and important are some other packings that are more or less random [Torquato and Stillinger (2010)]. An ordered packing of identical hard spheres could serve as a model for the crystal structures of solids, whereas its random counterparts could play the same role for the disordered structures of liquids, as was first proposed by John Desmond Bernal [Bernal (1960); Finney (2013)].

Random close packings of spheres can be generated if the spheres are poured randomly and then shaken to settle in a denser, jammed configuration. Empirical studies suggest that the maximum possible packing fraction for random close packings of identical spheres is only about 64%. Being substantially smaller than the value 74.048% for the densest ordered packings of FCC or HCP, this result clearly demonstrates the constraint of structural randomness on the space-filling ability of spheres.

Is there a way to improve this figure for random close packing? Yes, but it requires a different particle shape. In 2004, Paul Chaikin, then at Princeton University, found that the maximum random packing fraction for M&Ms (oblate spheroids, in mathematical parlance) is around 68%, surprisingly greater than that for identical spheres [Donev *et al.* (2004b)]. Similar results hold in two dimensions, where random packings of ellipses may be denser than their counterparts with circles [Delaney *et al.* (2005)].

How can Chaikin's results be understood? Imagine if you started with a box of randomly close-packed Maltesers (a spherical candy) and could somehow shrink the whole pack—candy and all—along one direction, until it had turned into a pack of M&Ms (i.e., oblate spheroids), with all of them aligned in the same orientation. The packing fraction would remain the same at 64% because the shrinking transformation scales the candy and the empty space equally. If you now shake this new pack repeatedly, the M&Ms can shift into different orientations, allowing them to fill space more efficiently. It may seem especially obvious if you imagined doing the

Figure 1.2 Molecular arrangements of certain liquid crystals with banana-shaped molecules [Takezoe and Takanishi (2006)].

shrinking horizontally, so that the M&Ms were all perched on their edges before you shook them.

1.5 Applications in the Physical Sciences

Some non-spherical particles also serve as a model for liquid crystals, the class of materials well-known for its role in liquid crystal displays (LCD) such as those in your wristwatch and mobile phone. Many liquid crystals are based on rod- or banana-shaped molecules. Our understanding of the partially ordered, liquid crystalline states of such materials is often based on descriptions of how such model particles are packed (see Figure 1.2) [Takezoe and Takanishi (2006)].

Packing problems are also relevant to information theory. Here, the sphere-packing problem in various dimensions relates to the densest packing of encoded information in a way that also reduces the risk of coding errors [Thompson (1983)]. For instance, the problem of packing 8-dimensional spheres in 8-dimensional space, where each sphere can be mapped to a byte (8 bits) of information, suddenly turns from something esoteric into a technology-driven requirement.

1.6 Packing Problems as a Growing Research Field

Due to their cross-disciplinary nature and wide scope, packing problems have over the years become a subject of study in their own right [Rogers (1964); Conway and Sloane (1999); Aste and Weaire (2008)], both as pure mathematical problems (e.g., proof of the Kepler conjecture [Hales (2005)]) and in some more applied contexts (e.g., the structure of emulsions [Jorjadze *et al.* (2011)]). To provide a platform for packing researchers from different disciplines to disseminate their results and to foster the development of such a research community, the Foams and Complex Systems research group of Trinity College Dublin organized a first ever International Workshop on Packing Problems in 2012, which had managed to attract speakers and participants from a range of disciplines including physics, mathematics, computer science and engineering. This was followed by subsequent packing conferences in Germany (Erlangen 2014), China (Shanghai 2016), and the United States (New Haven 2019), respectively.

Great progress on packing problems has been made via computational approaches such as simulated annealing [Pickett *et al.* (2000); Mughal *et al.* (2011, 2012)], linear programming [Fu *et al.* (2016, 2017)], molecular dynamics simulations [Koga and Tanaka (2006)], Monte Carlo simulations [Duran-Olivencia and Gordillo (2009); Jin *et al.* (2020); Jin *et al.* (2021)], and sequential deposition [Chan (2011, 2013); Chan *et al.* (2019)], leading to insights into possible theoretical solutions. However, there remain many circumstances where a combination of novel algorithms, greater computer power, and new mathematical theories is much needed. Some more complicated packing problems, such as those with deformable or complex-shaped objects (e.g., the biological packing of RNA during virus formation [Borodavka *et al.* (2012)]) particularly need new methods. But so do some centuries-old packing problems (e.g., the Kelvin problem [Weaire (1996)]) that are still waiting to be solved.

Chapter 2

An Introduction to Columnar Structures

This chapter presents an overview of what columnar structures are, how they can be classified via the phyllotactic notation, their applications in different contexts (e.g., botany, foams, and nanoscience), and how they can be obtained computationally via some generic models. It serves as a prelude to Chapter 3, which includes a detailed discussion of the models and concepts for the study of columnar structures.

2.1 A Friday-Afternoon Experiment: Packing Golf Balls into a Tube

It was a typically cold and rainy Friday afternoon in Dublin, Ireland. The weather was so depressing that we did not feel like doing anything else but an experiment involving golf balls and an acrylic tube: We dropped a total of approximately 40 golf balls sequentially

Columnar Structures of Spheres: Fundamentals and Applications
Jens Winkelmann and Ho-Kei Chan

ISBN 978-981-4669-48-1 (Hardcover), 978-0-429-09211-4 (eBook)
www.jennystanford.com

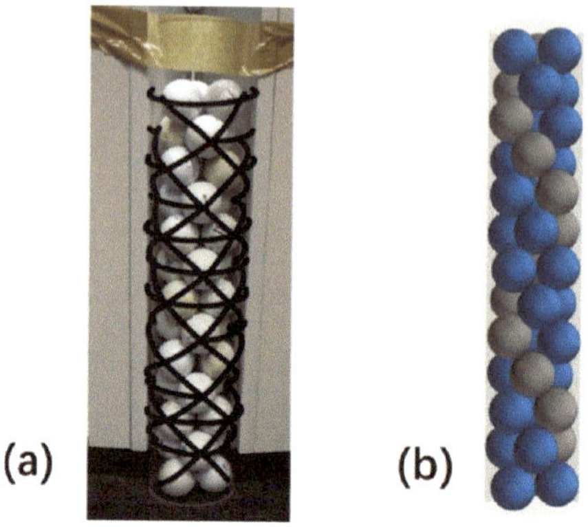

Figure 2.1 (a) An illustrative experimental realisation of a columnar structure of golf balls, and (b) an illustration of the corresponding simulated structure. Such a structure comes into existence when golf balls are dropped sequentially into a tube with a diameter ratio of $D \equiv D_{\text{cylinder}}/d_{\text{sphere}} \approx 2.22$ between the diameter D_{cylinder} of the cylindrical tube and the diameter d_{sphere} of the golf balls. The black lines indicate the contacts between adjacent golf balls. A short video of the full assembly process can be found in Reddit [Winkelmann (2019)] or Wikipedia [Winkelmann (2021)].

into the transparent tube, of which the diameter was a little bit larger than twice the diameter of each golf ball. To our surprise they stacked up to a perfectly ordered columnar structure, in which the golf balls are densely packed without any defects.[1]

This simple experiment of packing golf balls into a cylindrical tube is one illustrative method of creating what we refer to as *columnar structures*. The resulting columnar structure of this Friday-afternoon experiment is displayed in Figure 2.1(a). The first three golf balls at the bottom form an equilateral triangle. Any subsequent golf ball ends up located at the valley of two existing balls that lie adjacent to each other, when being dropped sequentially. This results in a hexagonal-like contact network as highlighted by the black lines. This particular structure differs from a perfect hexagonal packing by the absence of specific contacts between adjacent spheres. This exceptional type of structures, as introduced in Subsection 3.3.2, is

[1] Links to the video [Winkelmann (2019, 2021)]:
`https://www.reddit.com/r/ScienceGIFs/comments/b2c604/ordered_columnar_structures_golf_balls_packed/` and
`https://en.wikipedia.org/wiki/Cylinder_sphere_packing`

a particular focus of this monograph. As illustrated in Figure 2.1(b), such structures can also be obtained from computer simulations. Various methods for such simulations are discussed in detail in Chapter 3.

The resulting close-packed structure of golf balls in the above example depends on a number of factors related to the mechanics and dynamics of the system. These include friction, assembly processes, and the history of the structure. If we neglect such factors for the sake of simplicity, this packing problem can be reduced to a pure geometrical problem. The only control parameter for such a simplified model is the diameter ratio $D \equiv D_{\text{cylinder}}/d_{\text{sphere}}$, where D_{cylinder} and d_{sphere} are respectively the diameters of the cylindrical tube and the spherical golf balls. The rules of golf dictate the golf-ball diameter to be $d_{\text{sphere}} \approx 4.27$ cm, and the tube diameter is measured to be $D_{\text{cylinder}} = 9.50$ cm. We therefore estimate the diameter ratio for our experiment to be $D \approx 2.22$. At this diameter ratio, the same densest possible structure has been obtained from simulations [Figure 2.1(b)].

The mathematical problem of packing identical spheres into a cylinder is unique in terms of the spherical symmetry of the confined objects and the cylindrical symmetry of the confining space. The presence of these two types of symmetry simplify the corresponding computer simulations and structural analysis. For example, isotropic inter-particle interactions can be implemented in the simulations. For a specific range of the diameter ratio D, the spheres of any densest-packed structure are all in touch with the cylindrical boundary and therefore of the same radial distance from the long axis. This allows the packing problem to be treated as a two-dimensional one.

2.2 What Are Columnar Structures?

Any structure that constitutes densely packed objects of spherical-like shape in cylindrical confinement is hereafter referred to as a columnar structure of spheres. This broad definition includes random packings of spheres inside a cylindrical container, as well as ordered structures like the one shown in Figure 2.1. In this monograph, however, we are only concerned with columnar structures that are made up of equal-sized or monodisperse spheres (The adjectives "equal-sized" and "monodisperse" can be used interchangeably). In addition, we only consider columnar structures that are *ordered*, do not have point defects, and do not contain any *internal sphere*. All spheres of any columnar structure concerned here are in touch with the cylindrical wall, such that the columnar structure resembles the packing of disks on an unrolled cylindrical surface. Such structures exhibit a high degree of symmetry and can generally be simulated via twisted periodic boundary conditions. Ordered structures with all their spheres in touch with the cylindrical wall may still exhibit point defects. However, such structures are outside of the scope of this monograph. The nature of columnar structures within the scope of this monograph is explained in greater details in Section 3.4.

Packing equal-sized spheres into a cylinder (like our Friday-afternoon experiment as shown in Figure 2.1) is one simple way to obtain such ordered structures. However, such structures can also be obtained by packing spheres onto the surface of a cylinder or via a novel method involving rapid rotations [Lee *et al.* (2017)].

A. Rogava even managed to build such a structure without any confinement. He built a tower of tennis balls (bottom-left image of Figure 2.2) with a structure that resembles that of our golf-ball experiment [Rogava (2019)]. In his experiments, the friction between tennis balls is crucial for the mechanical equilibrium of each structure. Without such friction, there would be no torque balance and the structure would collapse.

He started his experiments by building a pyramid-like structure of tennis balls (upper-right image of Figure 2.2). By carefully removing some of the balls in the periphery, he ended up with a symmetric structure of sixteen balls (top-left image of Figure 2.2). In a second attempt, he was able to build a Christmas-tree-like structure with fourteen balls (middle-right image of Figure 2.2). Following a similar approach using more tennis balls, he ended up with a small columnar-like tower of seven tennis balls (bottom-right image of Figure 2.2). By adding more tennis balls to the tower, he obtained a structure that resembles our columnar structure of golf balls (bottom-left image of Figure 2.1). The top ball is crucial for the mechanical stability of the structure: through friction and gravity, it prevents the three balls beneath from falling apart.

Figure 2.2 Building columnar-like structures of tennis balls [Rogava (2019)]. By removing the peripheral tennis balls of a pyramid-like structure (top-right image), various types of columnar-like tennis-ball towers have been created. The one shown in the bottom-left image resembles the columnar structure shown in Figure 2.1. Each structure is maintained at mechanical equilibrium by friction and gravity. Figure reprinted, with permission, from Ref. [Rogava (2019)].

In this book, we focus on two different approaches of obtaining columnar structures of spheres. One method is the self-assembly of identical spheres inside a cylindrical tube, while the other (and more novel) method involves rapid rotations of spheres inside a liquid-filled tube. The latter method was first used by Lee *et al.* [Lee *et al.* (2017)] to create such structures experimentally. For this method, we describe a theoretical method that predicts the ordered structures to be obtained at given experimental conditions.

2.3 The Phyllotactic Notation: Categorising Columnar Structures

One way of classifying ordered columnar structures is based on the phyllotactic notation borrowed from botany. As its ancient Greek origin suggests (*phyllon* = leaf, *táxis* = arrangement), the term "phyllotaxis" is related to arrangements of leaves. The term was first coined by Charles Bonnet in the 18th century [Livio (2008)]. In his book "On Growth and Form" (1917), the mathematical biologist D'Arcy Thompson analysed such arrangements of plant parts around an axis [Thompson (1917)]. These include not only cylindrical structures like leaves around a stem, pine cones, or pineapples, but also the planar patterns of florets in a sunflower head. While the arrangements in the former are cylindrical, the spirals in the latter are arranged on a disk.

Phyllotaxis in the context of disk packings has recently been investigated by Sadoc *et al.* [Sadoc *et al.* (2012)]. They looked at the dense organisation of small disks inside a large circular domain. By using an algorithm that packs the disks into a Fermat spiral, they investigated the very nature of this assembly. Our focus, however, will be on phyllotaxis in the context of cylindrical structures.

Such a phyllotactic pattern can be obtained by rolling out the positions of the spheres within a columnar structure, as well as the corresponding sphere-sphere contacts, into a plane surface (see Figure 2.3). The centre of each sphere is displayed as a dot, and the contact between any pair of adjacent spheres is represented by a line.

In this monograph, we focus on columnar structures with phyllotactic patterns that resemble a hexagonal lattice, such as the one shown in Figure 2.3. Besides a perfectly hexagonal lattice, other types of phyllotactic patterns are also possible. For example, the rolled-out pattern for a linear chain of spheres (the simplest columnar structure) consists of straight lines only. Square-like patterns can be observed for zigzag structures as well as for columnar structures with internal spheres [Mughal *et al.* (2012)]. However, these are outside of the scope of this monograph.

Any phyllotactic pattern is characterised by a periodicity vector $\vec{V}$ that points from any sphere to its first periodic image along the angular dimension. This periodicity vector can be decomposed into

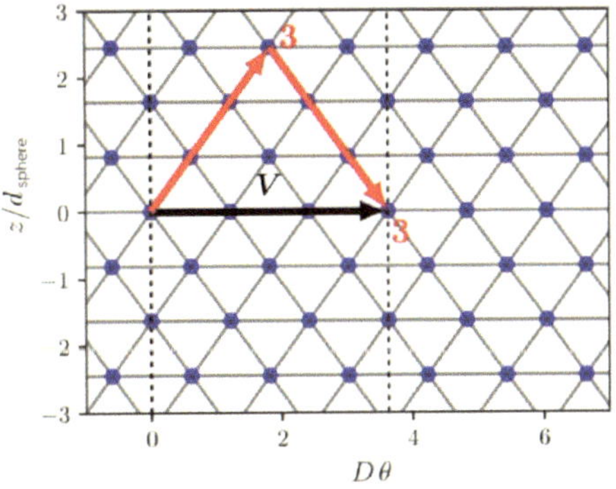

Figure 2.3 Rolled-out contact network of the $(3, 3, 0)$ structure. Each blue dot represents a sphere at some vertical position z and azimuthal angle θ, and each grey line corresponds to a contact between adjacent spheres. The magnitude of the periodicity vector $\vec{V}$ (black arrow) is equal to the circumference of the cylinder. This periodicity vector can be decomposed into components along any two of the three crystallographic directions (i.e., directions of the red or black arrows in the figure). In this example, the periodicity vector lies along one of the three crystallographic directions and is equal to three times of the corresponding basis vector, hence one duplet in the phyllotactic notation is $(3, 0)$. Alternatively, the periodicity vector can be viewed as a sum of vector components (red arrows) along the other two crystallographic directions, hence another duplet is $(3, 3)$. Combining these two duplets yields the phyllotactic notation $(3, 3, 0)$.

components along any two of the three crystallographic directions. To specify the three possible ways of doing so, a phyllotactic notation of three positive integers $(l = m + n, m, n)$ with $l \geq m \geq n$ is employed. For each crystallographic direction, the corresponding integer of the duplet denotes the magnitude of the vector component in units of the basis vector, which corresponds to the shortest distance between any two points along the chosen crystallographic direction.

For instance, Figure 2.3 displays the phyllotactic pattern for the $(3, 3, 0)$ structure. This structure is similar to the one shown in Figure 2.1; their differences are marginal. It consists of layers of spheres stacked on top of each other with the presence of three spheres in each layer. Each sphere is here in contact with all of its six adjacent spheres. The periodicity vector $\vec{V}$, which lies along one of the three crystallographic directions, wraps around the cylinder until it reaches the first periodic image of the original sphere. With this periodicity vector being three times of the corresponding basis vector, one duplet of integers in the phyllotactic notation is $(3, 0)$. On the other hand, this periodicity vector can be decomposed into components along the other two crystallographic directions (i.e., the red arrows in Figure 2.3), with each component being three times of the corresponding basis vector. Hence, another doublet of integers in the phyllotactic notation is $(3, 3)$. A unified description of these two duplets of integers yields the phyllotactic notation $(3, 3, 0)$.

The rolled-out contact network of our golf-ball packing in Figure 2.1 is the same as that shown in Figure 2.4. There exist missing contacts between adjacent spheres, as also observed for the highlighted contact network in Figure 2.1. Figure 2.4 reveals that such loss of contacts appear along a line, hence the structure is referred to as a line slip (see Subsection 3.3.2 for details). Since such missing contacts occur along either component of the periodicity vector $\vec{V}$, we mark one of the "3"s in the phyllotactic notation with a bold number, i.e., $(3, \mathbf{3}, 0)$.

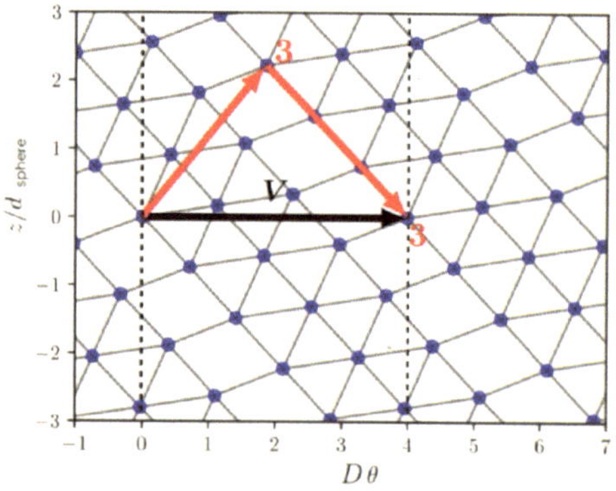

Figure 2.4 Rolled-out contact network of the columnar structure in Figure 2.1. Each blue dot represents a sphere at some vertical position z and azimuthal angle θ, and each grey line corresponds to a contact between adjacent spheres. This contact network consists of missing contacts that can also be seen in Figure 2.1. The structure is classified as a $(3, \mathbf{3}, 0)$ line slip in the phyllotactic notation.

2.4 Applications of Columnar Structures: From Botany and Foams to Nanoscience

Ordered columnar structures appear in different contexts on a broad range of length scales from the macro- to the nano-scale. Figure 2.5 illustrates examples from four different contexts ordered by their length scales. On the macroscopic scale, such structures can be found in botany where the seeds of a plant self-assemble around the stem. Plants like the *titan arum* can be up to a metre in size [www.cambridge2000.com (2005)].

At a similar scale, approximately equal-sized bubbles crystallise into an ordered columnar structure when confined in a glass tube. The diameters of both the confined bubbles and the confining tube are typically of the order of centimetres.

In nanoscience, such columnar structures can be found in synthetic materials that range from the micro- to the nano-scale. In some novel rod-like or fibre-like materials, the confined spherical

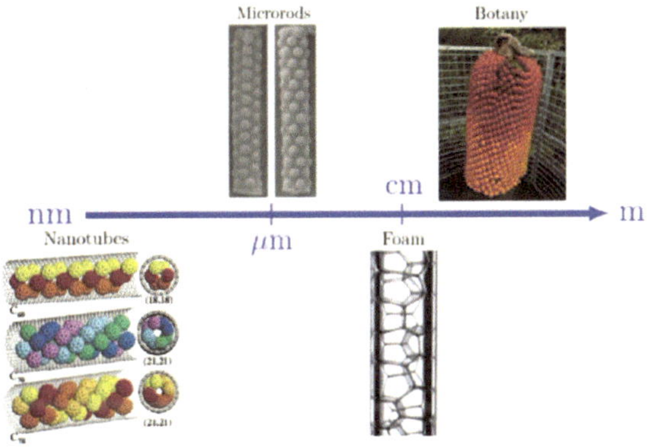

Figure 2.5 Examples of columnar structures ranging from nanometres (nm) to metres (m) in size. The example at the nano-scale corresponds to nanotube-confined particles (Figure reprinted, with permission, from Ref. [Troche *et al.* (2005)]). They can also be used to build rods/fibres where the rods/fibres are of micron size (Figure reprinted, with permission, from Ref. [Wu *et al.* (2017)]). Examples at the macro-scale involves columnar structures in foams and botany. The plant shown here is the *titan arum* (also called corpse flower) (Figure reprinted, with permission, from Ref. [www.cambridge2000.com (2005)]). All examples are discussed in detail in the main text.

entities are at the nano-scale, but the length of the confining cylindrical space is at the micro-scale [Wu *et al.* (2017); Liang *et al.* (2018)].

Furthermore, such columnar structures also show up when molecules such as fullerenes (C60, C70 or C78) are injected into nanotubes [Troche *et al.* (2005)]. In such cases, the confined particles can be used to alter the chemical and physical properties of the confining nanotubes.

In this section, we discuss all these examples from different length scales in detail. We first discuss two examples from botany as well as a selected range of botany-related literature. We then discuss some relevant research on foams and some novel applications from nano-science. Wu *et al.* [Wu *et al.* (2017)] from Penn State University built microrods of Si nanoparticles, which display structures that are similar to those obtained from our simulations. At Harvard, Tanjeem *et al.* [Tanjeem (2020)] investigated candidates of optical metamaterials, which are materials with a negative refractive index.

2.4.1 Examples from Botany

Because of their appearance in plants, columnar structures were first studied in botany [Thompson (1917); Erickson (1973)]. But they are also of interest in other aspects of biology [Bryan (1974); Hull (1976); Brinkley (1997); Amir and Nelson (2012)], such as bacteria, viruses, and microtubules. In 2018, such structures were discovered in the notochord of the zebra fish as well [Norman *et al.* (2018)].

Much research on the ordered arrangements of lateral organs in plants has been carried out by biologists and mathematicians alike. Their main interests lie in flowers and fruits that exhibit such spiral phyllotactic patterns. Reviews on this are given by Erickson [Erickson (1983)] and by Prusinkiewicz and Lindenmeyer [Prusinkiewicz and Lindenmeyer (1990)]. Most of the mathematical models relate such ordered arrangements in plants to packing problems. Prusinkiewicz and Lindenmeyer [Prusinkiewicz and Lindenmeyer (1990)] proposed a model that reduces phyllotaxis to the problem of packing circles on the surface of a cylinder. In their model, the arrangement of lateral organs in a plant is described

(a)  (b) 

Figure 2.6 Two examples of columnar structures from botany. The berries or seeds self-assemble around their stem in a columnar fashion. Figure (a) displays an *arum maculatum* from Bushy Park, Rathfarnham, Dublin, and Figure (b) displays an Australian bottlebrush from Templeogue, Dublin.

purely by geometry, similar to the models that we use throughout this research.

One of the largest flowers where the berries arrange in a regular cylindrical form is the *titan arum*, also referred to as "corpse flower" due to its rotten smell. An image of it is displayed at the top right of Figure 2.5. This flower can be of up to 3 m height and is natively and solely found in western Sumatra and western Java.

Some less exotic plants with such structures can also be found in Ireland. Two examples from Dublin are presented in Figure 2.6. Figure 2.6(a) displays a plant called "arum maculatum", which is commonly known as "jack in the pulpit" or "cuckoo-pint". Its height is approximately 20 cm and its berries are similar to those of the corpse flower, which is its larger relative. For this species, the spacial arrangement of the berries generally varies with the stem-to-berry size ratio.

Another species that can be found in many gardens of residential areas in Ireland is the Australian bottlebrush (see Figure 2.6(b)), where its seed capsules are assembled around the branches of the species.

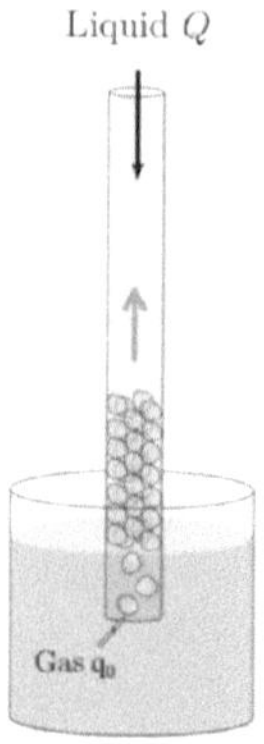

Figure 2.7 Experimental set-up to produce columnar foam structures. Gas is introduced at a constant flow rate q_0 into a surfactant solution to create bubbles of equal size. Using a bubbling needle, the bubbles of equal size are filled sequentially into a glass tube from the bottom. There they crystallise into a columnar foam structure. Their liquid fraction is increased by feeding the foam column with a surfactant solution at a liquid flow rate Q from the top (forced drainage).

2.4.2 Dry and Wet Foam Structures

Other macroscopic examples of ordered columnar arrangements include foam bubbles confined inside a glass tube. They can be realised experimentally by injecting equal-sized soap bubbles into a glass tube [Pittet *et al.* (1995); Saadatfar *et al.* (2008); Tobin *et al.* (2011); Meagher *et al.* (2015)]. In Section 5.3, we discuss various types of experimentally obtained foam structures. We then compare these structures with those obtained from computer simulations. The corresponding experimental set-up, as well as some important findings in recent years, are also discussed in this subsection.

Columnar structures of foams can be created experimentally by filling the tube sequentially with bubbles from the bottom (see Figure 2.7 for an illustration of the experimental set-up). A steady stream of bubbles is produced by blowing air through a needle dipped in a surfactant solution. Bubbles of equal size are created at a constant gas-flow rate q_0. They are then collected in a cylindrical tube, where they crystallise into a columnar foam. The resulting foam column is put under forced drainage by feeding it with surfactant solution from the top at some liquid-flow rate Q. Using this method, the amount of liquid (or the liquid fraction) in the foam can be adjusted [Hutzler *et al.* (1997); Weaire *et al.* (1997)].

Depending on the liquid fraction in the foam, the columnar foam crystal can either be *dry* or *wet*. In a dry foam, the liquid fraction is low and each cell appears to be a polyhedron (see the foam structures in Figure 2.5 and Figure 2.8). In a wet foam, each bubble takes on a spherical shape as a result of a high liquid fraction [see Figure 2.8(c)].

Columnar crystals of foams from the dry to the wet limit have been extensively studied over the last 20 years [Pittet *et al.* (1995); Hutzler *et al.* (1997, 1998); Saadatfar *et al.* (2008); Tobin *et al.* (2011); Meagher *et al.* (2015)]. Experimentally discovered foam structures as reported in the literature [Tobin *et al.* (2011); Meagher *et al.* (2015)] exhibit a close structural resemblance with the sphere packings presented in this monograph. A variety of such foam structures have been discovered and then classified in terms of the phyllotactic notation.

Previous work on columnar foams has focussed not only on experiments, but also on computer simulations. Most simulations in the past have been carried out using the Surface Evolver [Brakke (1992)] software and have focused mainly on dry structures [Drenckhan *et al.* (2005); Saadatfar *et al.* (2008); Tobin *et al.* (2011)]. Surface Evolver is a computer program for modelling surfaces shaped by surface tension or other types of constraints. It is based on the concept of surface-energy minimisation, i.e., it evolves a given surface structure towards minimal energy by various types of numerical minimisation routines. Figure 2.8(a) shows a zigzag structure as obtained from the Surface Evolver, and Figure 2.8(b) displays an experimental image of this structure [Tobin *et al.* (2011)].

This simple zigzag structure has been a subject of many experimental investigations. The observation of a moving interface with increasing liquid fraction was reported by [Hutzler *et al.* (1997)]. This includes an unexpected 180° twist interface, an explanation of which is still lacking.

Other discoveries include complex structures with internal spheres (or foam cells) [Saadatfar *et al.* (2008); Meagher *et al.* (2015)]. Some dry-foam structures with interior cells were found to consist of a chain of pentagonal dodecahedra or Kelvin cells in the centre of the tube [Saadatfar *et al.* (2008)]. Many more arrangements of this type were also investigated in detail, for example by X-ray

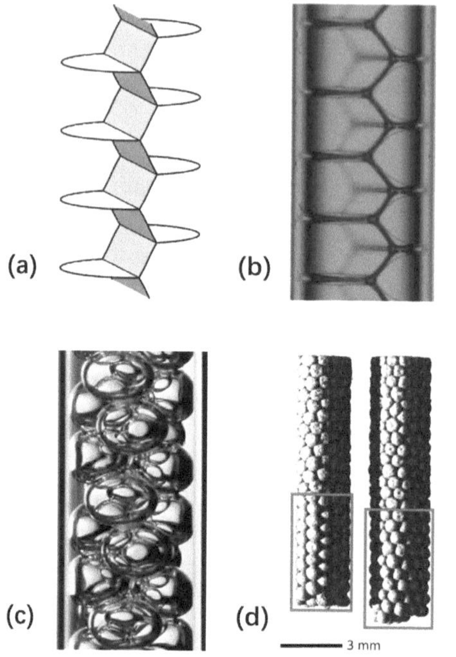

Figure 2.8 Various types of columnar structures of soap bubbles confined inside a glass tube. Figures (a) and (b) show a zigzag structure of a dry foam in Surface Evolver and experiments, respectively. Both figures reprinted, with permission, from Ref. [Tobin *et al.* (2011)]. Figure (c) displays a wet foam from a forced-drainage experiment. In Figure (d), some more complex foam structures with internal spheres are shown. These structures were imaged by Meagher *et al.* using X-ray tomography (Figure reprinted, with permission, from Ref. [Meagher *et al.* (2015)]).

tomography for an accurate determination of the position of each bubble [Meagher *et al.* (2015)]. It was observed that the outer bubble layer is ordered, with each internal layer resembling a different, simpler columnar structure. Two examples imaged with X-ray tomography are presented in Figure 2.8(d).

Transitions between different foam structures were also explored experimentally by two different techniques: The bubble diameter can be altered by adjusting the gas supply that creates the bubbles, changing the effective tube-to-bubble diameter ratio. Structures can also be transformed by dilating or compressing the

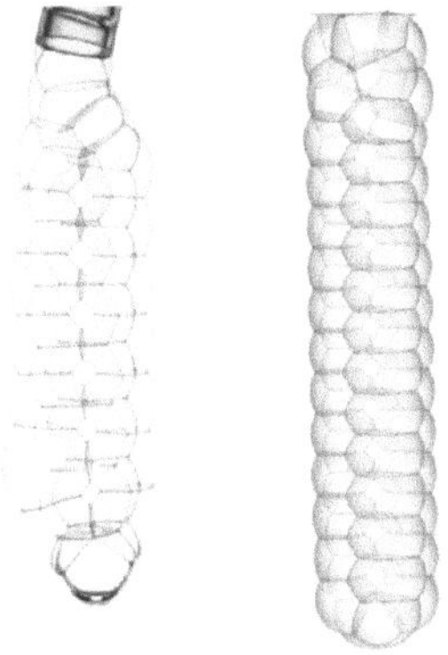

Figure 2.9 Experimental image taken by the photographer K. Cox (left), and Surface Evolver simulations of recently discovered chain-like arrangements of soap bubbles (right) [Cox *et al.* (In preparation)]. In the experiments, monodisperse bubbles were slowly pushed out of a vertical tube, where they spontaneously self-assemble into a columnar structure. In the Surface Evolver simulations, gravity was implemented to preserve the property of cylindrical symmetry.

foam inside the tube. Pittet and Boltenhagen used both techniques to enforce such crystallographic transitions in Surface Evolver simulations as well as in experiments [Pittet *et al.* (1995, 1996); Boltenhagen and Pittet (1998); Boltenhagen *et al.* (1998)].

K. Cox recently discovered that such foam structures can also arise from a chain-like arrangement of soap bubbles without the need of cylindrical confinement [Cox *et al.* (In preparation)]. The soap bubbles self-assemble into such structures when being pushed out of a vertical tube. There they spontaneously arrange into various types of columnar structures (see the example in Figure 2.9).

The image on the left of Figure 2.9 displays such a bubble chain taken by K. Cox. The same structure can be obtained computationally using Surface Evolver (image on the right of Figure 2.9). In the simulations, the bubble chain was not subject to any sort of cylindrical confinement. The top bubbles were fixed to a ceiling, and the property of cylindrical symmetry was preserved by stretching the bubble chain under gravity.

Preliminary experiments and simulations have shown that the structure of a short chain is solely determined by the corresponding bubble-to-tube diameter ratio. For a long chain, there exists a nontrivial pressure gradient as a result of gravity, and hence a structural variation along the chain is expected [Cox *et al.* (In preparation)].

2.4.3 Nanoscience: Microrods and Optical Metamaterials

Up to now we have discussed the ubiquitous existence of macroscale columnar structures in the contexts of botany and foams. Such structures have also been studied in the physical sciences at smaller scales, particularly in nanoscience [Brisson and Unwin (1984); Chopra *et al.* (1995); Hodak and Girifalco (2003); Smalley (2003); Khlobystov *et al.* (2004); Wu *et al.* (2004); Troche *et al.* (2005); Yamazaki *et al.* (2008); Tymczenko *et al.* (2008); Lohr *et al.* (2010); Warner and Wilson (2010); Legoas *et al.* (2011); Sanwaria *et al.* (2014); Wang *et al.* (2016); Wu *et al.* (2017); Liang *et al.* (2018); Tanjeem (2020); Zhang *et al.* (2019)], particularly in nanoscience [Chopra *et al.* (1995); Hodak and Girifalco (2003); Smalley (2003); Khlobystov *et al.* (2004); Wu *et al.* (2004); Troche *et al.* (2005); Yamazaki *et al.* (2008); Warner and Wilson (2010); Legoas *et al.* (2011); Sanwaria *et al.* (2014); Wang *et al.* (2016); Zhang *et al.* (2019)]. There they appear in a huge variety of synthetic materials. For example, they are popular candidates for the fabrication of nanowires, microrods, or microfibres, where the stiffness of such quasi-one-dimensional systems is determined by the corresponding microstructures [Wood (2017)].

Many researchers also tried to alter the properties of nanotubes by trapping identical particles inside them [Chopra *et al.* (1995); Hodak and Girifalco (2003); Smalley (2003); Khlobystov *et al.* (2004); Troche *et al.* (2005); Yamazaki *et al.* (2008); Warner and Wilson (2010); Legoas *et al.* (2011); Wang *et al.* (2016)]. These were mostly done via the self-assembly of fullerenes such as C60, C70, or C78 within carbon nanotubes (see the bottom-left image of Figure 2.5) [Hodak and Girifalco (2003); Khlobystov *et al.* (2004); Troche *et al.* (2005); Yamazaki *et al.* (2008); Warner and Wilson (2010)] or boron nitride nanotubes [Chopra *et al.* (1995)].

Figure 2.10 displays four images from different fields of nanoscience [Wu *et al.* (2017); Lázaro *et al.* (2018); Tanjeem (2020)]. In this subsection we will describe those examples in detail and provide some background information about the research. While

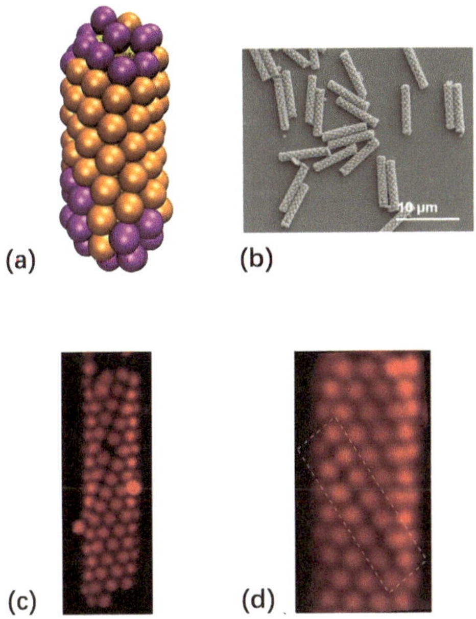

Figure 2.10 Examples of columnar structures in nanoscience. (a) Simulation of particles coated on the surface of a spherocylinder (length $L_{\text{cyl}} \sim 45$ nm) where they are used for drug delivery (Figure reprinted, with permission, from Ref. [Lázaro *et al.* (2018)]). (b) Microrods of length of a few microns can be created by depositing Si nanoparticles inside PDMS pores (Figure reprinted, with permission, from Ref. [Wu *et al.* (2017)]). In Figures (c) and (d), candidates of optical metamaterials are shown (Figures reprinted, with permission, from Ref. [Tanjeem (2020)]). This is a material with a negative refractive index [Tanjeem (2020)]. The structures were constructed via a self-assembly of nanocolloids ($d \sim 500$ nm) onto the surface of a cylinder. A defected structure consisting of a line slip with a kink [Figure (c)] as well as a line-slip structure [Figure (d)] were discovered. The subject of line-slip structures is discussed in detail in Section 5.3.

the previous examples from botany and foam research has little applications outside their domains, the examples from nanoscience could be relevant to the contexts of materials science, such as liquid crystals or optical metamaterials.

Columnar-like structures also occur in pharmaceutical research where drug particles are coated onto the surfaces of spherocylinders [Lázaro *et al.* (2018)] as densely as possible. Lazáro *et al.* examined the morphologies of virus capsid proteins self-assembled around metal nanorods [Lázaro *et al.* (2018)]. One of their simulated packings on a spherocylinder is displayed in Figure 2.10.

Wu *et al.* built rods [Figure 2.10(b)] of the size of several microns [scale given in Figure 2.10(b)]. These microrods were created by packing silica colloidal particles densely into cylindrical pores. By solidifying the assembled structures, the microrods were imaged via scanning electron microscopy (SEM).

The assembly process consists of three steps. First, charged silica nanoparticles of diameter $d_{\text{sphere}} \sim 500$ nm are deposited inside PDMS (polydimethylsiloxane) nanopores of diameter $D_{\text{cylinder}} =$ 1.8 μm and length $L = 8$ μm. Such particles are dispersed inside a photo-cross-linkable monomer (ETPTA) with a similar refractive index. Due to the Si particles' long-range repulsive interactions, they are referred to as "soft colloids". Their softness yields an effective diameter which is generally much larger than the actual diameter.

Inside the PDMS pores the particles self-assemble into ordered cylindrical structures. In order to obtain a variety of structures for the same pore diameter and particle diameter, the concentration of particles is varied across different pores at the same pore length. In our simulations (Chapter 5), this corresponds to a variation of the uniaxial pressure.

By UV-curing the samples in a second step, the ETPTA crosslinks, solidifying the dispersion inside the PDMS pores to a solid cylindrical rod. The assembled Si particles are locked into place within this cylindrical rod and the PDMS mold around those rods can be peeled off.

Since this leaves the columnar structure hidden inside the solidified ETPTA, one last step is needed to make the arrangement of Si particles visible again. The outside polymer layer is removed by using oxygen plasma. With direct imaging techniques such as SEM, Wu *et al.* took images of the microrods as those seen in Figure 2.10(b).

From a closer look at Figure 2.10(b), one can observe a few distinct structures of their microrods. By increasing the concentration of nanoparticles inside the pores from 15% to 40%, they discovered 8 dominant types of closed-packed structures of spheres. Figure 2.10(b) also shows the coexistence of multiple packings within one microrod. Similar structures with this particular feature are discussed in Chapter 6.

In order to demonstrate the ability to obtain a desired structure in a controlled way, Wu *et al.* compared their structures to results of Molecular Dynamics (MD) simulations. They employed a pairwise-additive, long-range repulsive inter-sphere interaction, which is derived from a simple Yukawa potential. Similarly, wall interactions were implemented for the cylindrical surface as well as for the boundaries at the top and the bottom. By varying the number of colloidal particles in the cylinder until a steady state is attained, they obtained the same structures as those observed in the experiments. The nanoparticle concentration in the simulations was slightly smaller than that in the experiments.

Such microrods, with their physical properties such as stiffness or electrical conductivity depending sensitively on the assembled columnar structures, find potential applications in the context of materials science, such as liquid crystals. In Subsection 3.3.2, the structural dependence of electrical conductivity for line-slip structures is discussed. In a liquid crystal, the percolation threshold (the lowest concentration at which a liquid crystal is conductive) is directly proportional to the conductivity of its constituents. The structures of the microrods therefore have a direct influence on this percolation threshold. Other possible applications of microrods with a tunable conductivity include organic solar cells [Hill *et al.* (2004)] and other optoelectronic devices [Wu *et al.* (2017)].

Tanjeem *et al.* investigated columnar structures with the long-term goal of developing novel optical metamaterials [Schade *et al.* (2013); Manoharan (2019)], i.e., materials with a negative refractive index. No material with such a property exists in nature, but they can be manufactured artificially [Shalaev (2007); Ozbay (2008)].

Refraction of a light beam at the interface to a medium with a negative refractive index leads to a negative angle of refraction. Because of this property optical metamaterials find interesting applications in super lenses or optical cloaking [Cai *et al.* (2007); Ozbay (2008)]. Super lenses have the ability to go beyond the diffraction limit and can thus be used for super-resolution imaging.

An optical metamaterial is composed of constituents that are smaller than the wavelength of light [Manoharan (2019)]. For visible light that means the elements must be smaller than 100 nm. Most metamaterials of current research operate for microwaves, which have a wavelength on the microscale and thus feature larger micro-scale structures.

These constituents act like resonators that strongly scatter incoming light of a certain frequency. This light is scattered isotropically, which means that the resonance does not depend on the orientation of the resonator nor the direction of the incoming light [Manoharan (2019)].

Tanjeem *et al.* tried to construct such a resonator via the self-assembly of nanospheres onto the surface of a cylinder [Tanjeem (2020)]. The nanospheres ($d \sim 700$ nm) were therefore suspended in an SDS (sodium dodecyl sulfate) solution together with a cylinder of diameter D_{cylinder}. A depletion force between the nanospheres and the cylinders adheres the spheres to the cylinder. For this to take place, the cylinder diameter has to be much larger than the diameter of the nanospheres ($D \approx 3 - 5$).

Tanjeem *et al.* have obtained a variety of crystalline structures with some of them being metastable [Tanjeem (2020)]. Figure 2.10(c) and Figure 2.10(d) present respectively two example structures at the same value of D. The one shown in Figure 2.10(d) is a ground-state line slip with one straight helical line. The features of this type of structure are discussed in detail in Subsection 3.3.2.

Furthermore, Tanjeem *et al.* have discovered a new type of line-slip structures that contain kinks [Tanjeem (2020)], as illustrated in Figure 2.10(c). Upon a comparison with finite-temperature simulations, they found that the kinks correspond to low-energy excitations from line-slip structures [Tanjeem (2020)].

Due to their chirality, some columnar structures as obtained from nano-composites play an important role in photonics [Fan and Govorov (2011); Cong *et al.* (2014)]. Chiral structures include those that resemble a cork screw, such as the structure shown in Figure 2.10(c), whereas achiral structures are those that are symmetric around a rotational axis. The topic of chirality in columnar structures is discussed in more detail in Section 3.3.

Chiral structures possess some interesting optical properties that can be used for applications such as optical sensors or photonic crystals [Hill *et al.* (2004)]. For instance, chirality is a necessary geometrical feature for building a beamsplitter that splits incoming light into s- and p-polarised light [Turner *et al.* (2013a,b)]. Turner *et al.* [Turner *et al.* (2013b)] built such an optical device from a novel photonic crystal with a chiral asymmetry.

Of further interest in the context of photonic crystals are structures with a 4-fold symmetry. Those structures exhibit a discrete rotational symmetry, i.e., each of such structures looks the same if it is rotated around the vertical axis by an angle of $\pi/2$. This particular type of symmetry has the special property of destroying circular dichroism [Saba *et al.* (2013)]. Circular dichroism describes the effect of a material that absorbs left- and right-handed light of a circular polarising light source to different extents. Saba *et al.* discovered that crystals with a 4-fold symmetry destroy this property, making this geometry an attractive design for photonic materials. Columnar structures with a phyllotactic notation of the form $(4n, 4n, 0)$, where $n \in \mathbb{N}$, possess such a 4-fold symmetry.

2.5 Advantages of Generic-Model Simulations

For all these applications, columnar structures have been studied extensively. Many experiments and simulations have been carried out to examine their occurrence in specific scientific contexts.

Previous computer simulations such as the MD simulations by Wu *et al.* [Wu *et al.* (2017)] replicated the full corresponding experimental process of columnar-structure formation. The full movement of each sphere during this process is simulated, which makes the simulations computationally intensive as well as complex.

Many previous simulations also adopt inter-particle interactions that apply only to specific types of experimental systems. Troche *et al.* [Troche *et al.* (2005)], for instance, used specific models to simulate the intermolecular interactions of fullerenes. This allows a close comparison with corresponding experiments, but it also constraints the simulation results to a particular experimental context.

The research discussed in this monograph involved simulations that were based on generic models. While such models are only first-order approximations for the corresponding inter-particle interactions, they still yield a good agreement between simulations and experiments for columnar structures in mechanical equilibrium. Such concepts and generic models are introduced in Section 3.2.

Such simulations are computationally efficient because they only involve a search for structures in mechanical equilibrium without considering the trajectory of each sphere. Within relatively short computational times, important information about the equilibrium structures can be obtained. The simulation results can then be compared with experimentally observed structures or with structures obtained from some more complex simulation algorithms.

In some cases, the simplicity of a generic model allows one to obtain the mechanically equilibrated columnar structures through analytic derivations. One advantage is that no computational power is required to obtain information about the equilibrium structure. In Section 6.3, analytic results for columnar structures assembled in a process involving rapid rotations are presented.

The efficiency of computer simulations [Chapter 5] and the transparency of analytic derivation [Chapter 6] cater for the construction of phase diagrams and stability diagrams for different types of assembly methods. Such diagrams contain comprehensive information about the existence and stability of corresponding columnar structures, respectively.

The concepts and models for the study of columnar structures are introduced in Chapters 3 and 4, where in the latter a method of sequential deposition for constructing columnar structures of spheres is described in detail. In Chapter 5, we present some numerical as well as experimental results for soap bubbles inside a glass tube. In Chapter 6, we present a phase diagram that was obtained from the analytic calculations for a novel assembly process of columnar structures involving rapid rotations. Chapter 7 focusses on the equilibrium configurations of spheres inside a cylindrical harmonic potential. These have been investigated experimentally, as well as computationally via a generic model. In Chapter 8, we summarise all the findings discussed in Chapters 4 to 7 and present an outlook for possible future research.

Chapter 3

Models and Concepts for Columnar Structures

This chapter introduces the models and concepts that are employed to describe columnar structures of spheres. It introduces the reader to some important terminology used in subsequent chapters, and it discusses the theoretical models employed in our simulations of hard- and soft-sphere packings. We start with an introduction of the concept of packing fraction, followed by a discussion of the models for hard- and soft-sphere packings. We then move on to discuss the various types of structures, including uniform and line-slip structures, that have been observed for columnar packings of hard spheres. We review some previous work [Mughal et al. *(2012); Fu* et al. *(2016); Fu (2017); Fu* et al. *(2017)] on densest packings of hard spheres where uniform and line-slip structures and their structural transitions have been studied. Since most of our simulations are based on minimisation algorithms, the most common types of such algorithms are reviewed in this chapter.*

3.1 The Packing Fraction ϕ

An important physical property of any dense packing of spheres is its packing fraction ϕ. It is defined as the ratio of the total volume of a

Columnar Structures of Spheres: Fundamentals and Applications
Jens Winkelmann and Ho-Kei Chan

ISBN 978-981-4669-48-1 (Hardcover), 978-0-429-09211-4 (eBook)
www.jennystanford.com

set of objects to the volume of space into which they are packed. It is a measure of how efficiently the objects are packed into the given space.

In crystallography, the packing fraction has been well studied for various types of lattice packings of equal-sized spheres [face-centred cubic (fcc), body-centred cubic (bcc), and hexagonal close packed (hcp)]. Hales [Hales (2005)] proved that the highest packing fraction that can be achieved by any ordered packing of equal-sized spheres in three-dimensional space is $\pi/\sqrt{18} \approx 0.74048$, i.e., the Kepler conjecture. Any sphere packing arranged as a fcc or a hcp lattice achieves this packing fraction. The packing fraction of a bcc lattice is given by $\phi_{\text{bcc}} \approx 0.68$. The packing fractions of some other types of lattices have been investigated by O'Keeffe [O'Keeffe (1998)] and by Koch and Fischer [Koch and Fischer (2004)].

Let there be N spheres in the unit cell of an ordered columnar structures of spheres. This unit cell is the smallest repeating pattern of the entire ordered structure. The total volume of spheres in the unit cell is given by $NV_{\text{sphere}} = N(4/3)\pi(d_{\text{sphere}}/2)^3$, where d_{sphere} is the diameter of a sphere. The volume of the unit cell is given by $V_{\text{cell}} = L\pi(D_{\text{cylinder}}/2)^2$, where L and D_{cylinder} are respectively the length and diameter of the unit cell. The packing fraction is then given by

$$\phi = \frac{NV_{\text{sphere}}}{V_{\text{cell}}} = \frac{2Nd^3_{\text{sphere}}}{3LD^2_{\text{cylinder}}} = \frac{2}{3D^2}\frac{N}{L/d_{\text{sphere}}} \tag{3.1}$$

where $D \equiv D_{\text{cylinder}}/d_{\text{sphere}}$ is again the diameter ratio between the confining cylinder and the confined spheres.

3.2 Hard Spheres vs. Soft Spheres

As discussed in Chapter 2, the spherical constituents of columnar structures in the physical world can vary from golf balls and tennis balls to berries, seeds, soap bubbles, emulsion droplets, as well as micro- and nanospheres. While these various types of constituents share the same spherical geometry, they interact in many different ways. Soap bubbles, for instance, can be deformed when they come in contact with each other (as discussed further in Chapter 8).

Micro- and nanospheres are often described by the Lennard–Jones potential, which consists of an attractive and a repulsive term [Wood *et al.* (2013); Wu *et al.* (2017)].

However, in many cases, it is sufficient to understand, qualitatively and semi-quantitatively, the corresponding formation of columnar structures via a generic theoretical model. In this monograph, we focus on two generic models that are both based on repulsive inter-particle interactions: the *hard-sphere* model and the *soft-sphere* model. With purely repulsive interactions, we are disregarding any possible attractive electrostatic interaction between spheres.

Due to their simplicity, these models are relatively easy to implement in simulations. The hard-sphere model has already been used extensively for simulating columnar structures [Stoyan and Yaskov (2010); Chan (2011); Mughal *et al.* (2011, 2012); Mughal (2013); Fu *et al.* (2016); Fu (2017); Fu *et al.* (2017)]. For soap bubbles and colloids, the soft-sphere model is good enough for generating the corresponding columnar structures to a first order approximation.

3.2.1 The Hard-Sphere Model

The model of hard-sphere interactions originated from research on the statistical mechanics of fluids and gases [Hansen and McDonald (1990); Schroeder-Turk *et al.* (2015)]. There, the thermodynamic properties of hard-sphere systems have been investigated analytically as well as computationally.

The interaction of a pair of hard spheres is illustrated in Figure 3.1. The centre-to-centre separation between any two hard spheres cannot be smaller than the sum of their radii. For any pair of equal-sized spheres, their smallest centre-to-centre separation is equal to the diameter d_{sphere} of a sphere, i.e., the spheres are 'impenetrable' and do not overlap. Their interaction energy is binary: If the centre-to-centre separation between two hard spheres is greater than the sum of their radii, the energy is zero, otherwise it is infinite.

In Chapter 7, we use this model to describe equilibrium configurations of spheres inside a cylindrical harmonic potential. The numerical results are then compared against experimental results of

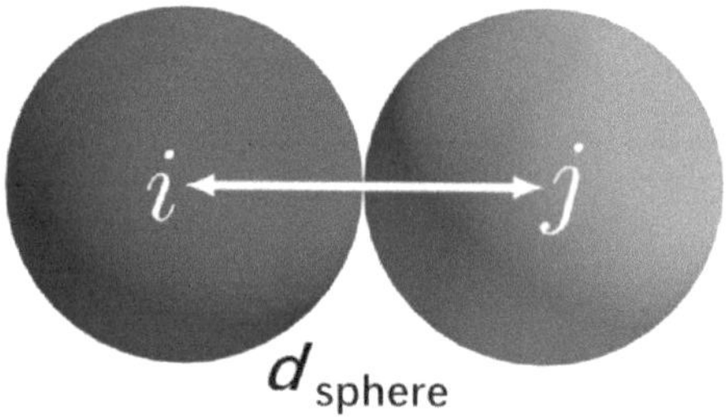

Figure 3.1 Illustration of the hard-sphere model: Hard spheres do not overlap. The centre-to-centre separation between two hard spheres i and j of equal size is always no less than the diameter d_{sphere}.

polymeric beads, which (as we will see) are impenetrable and thus behave like hard spheres.

3.2.2 The Soft-Sphere Model

Apart from columnar structures of hard spheres, theoretical studies on the columnar structures of soft, deformable spheres have also been carried out (Chapters 5 and 6). Such structures of soft spheres have been observed in foams and emulsions, as well as in colloids and nanoparticles with soft repulsive interactions [Lohr *et al.* (2010); Wood *et al.* (2013); Wu *et al.* (2017)].

In contrast to the hard-sphere model, the centre-to-centre separation between two soft spheres can be smaller than the sum of their radii, hence a possibility of overlapping between the particles. We define the overlap δ_{ij} for two spheres of the same diameter d_{sphere} as

$$\delta_{ij} = d_{\text{sphere}} - |\vec{r}_i - \vec{r}_j|, \tag{3.2}$$

where $\vec{r}_i$ and $\vec{r}_j$ denote the centre positions of spheres i and j, respectively.

Two soft (monodisperse) objects, such as two bubbles, deform when their centre-to-centre separation is smaller than their diameter

d_{sphere}. An interaction energy can be ascribed to such deformation. According to the soft-sphere model, the interaction energy E_{ij}^{S} between spheres i and j is a function of the square of the overlap δ_{ij} if the spheres are overlapping, and this energy is zero if there is no overlap between the spheres. This is mathematically formulated as

$$E_{ij}^{\mathrm{S}} = \begin{cases} 0 & \delta_{ij} > 0 \\ \frac{1}{2}k\delta_{ij}^2 & \delta_{ij} \leq 0 \end{cases} \tag{3.3}$$

where k is a constant.

This is the same energy expression as that in Hooke's law for a repelling harmonic spring, where k is the spring constant. In other words, two overlapping soft spheres repel each other like a harmonic spring, as illustrated in Figure 3.2(a). Such overlapping is penalised

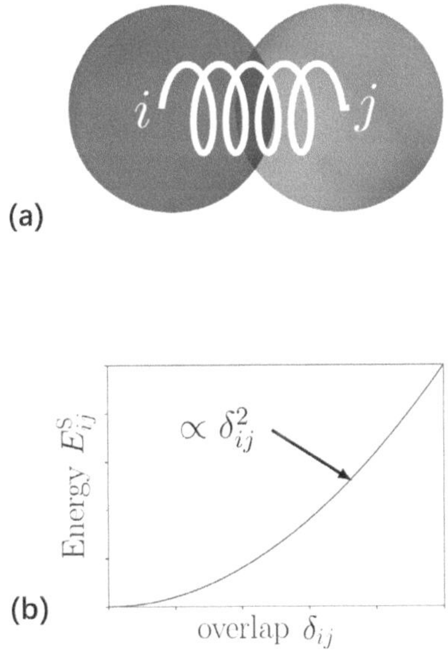

Figure 3.2 Illustration of a soft-sphere interaction. (a): Two soft spheres i and j interact like a repelling spring, when they exhibit an overlap δ_{ij} as defined by Equation (3.2). (b): Their interaction energy E_{ij}^{S} varies with the square of the overlap, i.e., $E_{ij}^{\mathrm{S}} \propto \delta_{ij}^2$.

with an interaction energy that scales with the square of the overlap [Figure 3.2(b)].

The softness of the spheres is modelled by the spring constant k. Results from the hard-sphere model can be recovered in the limiting case of $k \to \infty$. That is, our soft-sphere simulations are validated by an agreement between soft-sphere results in the hard-sphere limit and previous simulation results of hard-sphere packings [Mughal *et al.* (2012)].

With the consideration of only pairwise interactions, the soft-sphere model can be implemented with a high computational efficiency. For foams and emulsions, this model has been the most frequently employed model since it was proposed by Durian in 1995 [Durian (1995)]. Due to its good agreement with Herschel-Bulkley-type rheology in foams and emulsions, it is widely believed that the model is a good representation of such soft matter systems.

Similar to the Hertzian contact model [Johnson (1985)] (which describes contacts of elastic solids), the soft-sphere model describes contacts between bubbles or droplets based on similar assumptions as the Hertzian model. For both models, the strains are within the elastic limit, and the area of contact is much smaller than the cross-sectional area of a sphere. Any type of friction between the surfaces of spheres are generally neglected. However, sliding friction can be added to the soft-sphere model via the introduction of a pairwise frictional force between spheres in contact. This can only be implemented computationally via molecular dynamics (MD) simulations. We, however, adopt an (energy) minimisation approach that does not consider such sliding friction. For systems of soap bubbles or emulsion droplets, this is a valid approximation.

However, as a drawback, the soft-sphere model lacks volume conservation, since any inter-sphere overlapping in the model reduces the effective volumes of the spheres. In reality, when a bubble or an emulsion droplet is compressed, it bulges sideways such that the volume of any bubble or droplet is conserved to a first approximation. Furthermore, the pairwise interaction as used in the soft-sphere model is not a very accurate description for foam bubbles [Morse and Witten (1993); Lacasse *et al.* (1996)]. The limitations of this model in regard to foams and emulsions are discussed in Chapter 8.

A great advantage of the soft-sphere model, over the Surface Evolver [Brakke (1992)], `plat` [Bolton (1996)] or the recently developed deformable particle model (DPM) [Boromand *et al.* (2018, 2019)], is its simplicity and computational efficiency. The other models, which describe bubbles in a foam by films or vertices, are usually computationally inefficient because of their high numbers of films and vertices. The soft-sphere model, on the other hand, simulates a foam by simply approximating each bubble as a sphere.

3.3 Different Types of Columnar Structures

In previous simulations, a variety of packings have been obtained for hard spheres of diameter d_{sphere} packed inside a cylinder of diameter D_{cylinder} [Mughal *et al.* (2011, 2012); Mughal (2013); Fu *et al.* (2016); Fu (2017); Fu *et al.* (2017)]. Some of these observed structures, such as helical defect packings [Yamchi and Bowles (2015)], are metastable. In this monograph, we focus only on dense ordered packings that are stable for at least one single value of the diameter ratio D.

Ordered columnar structures can be categorized in different possible ways, one of which is based on their chirality. Some of these structures are chiral, while others are achiral. Chiral structures exhibit screw symmetry (like a cork screw), i.e., formation of helicity around the vertical axis. Mathematically speaking, a chiral structure cannot be mapped onto their mirror images by rotation or translation alone, while achiral structures exhibit such symmetry. Examples of a chiral and an achiral structure are presented in Figure 3.5 of Section 3.4.

In this monograph, columnar structures are categorized into two types: *uniform* structures and *line-slip* structures.

3.3.1 What Is a *Uniform* Structure?

An important type of dense columnar structures are referred to as uniform structures. The features of one such structure are illustrated in Figure 3.3. The example presented here is a so-called $(3, 2, 1)$ uniform structure. The triplet of phyllotactic notation can

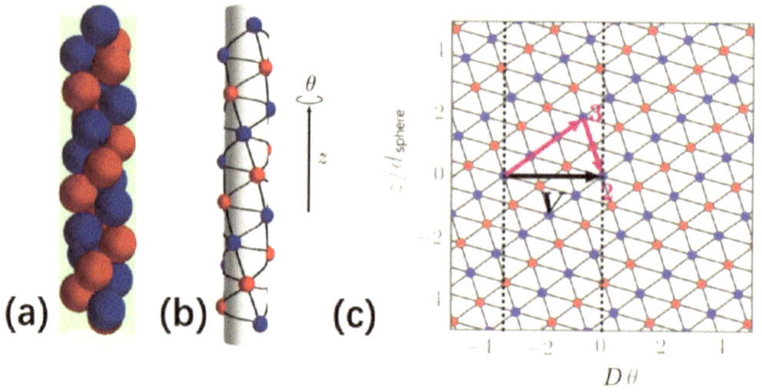

Figure 3.3 Illustration of a uniform structure. Figure (a) shows the three-dimensional packing of a $(3, 2, 1)$ uniform structure. Figure (b) displays the corresponding contact network, where each dot represents a sphere and each line corresponds to a contact between adjacent spheres. Figure (c) displays the contact network as rolled out into a plane of height z and azimuthal angle θ, showing a regular hexagonal lattice. From Figure (b) and Figure (c) it can be seen that every sphere is in contact with six other spheres. The periodicity vector $\vec{V}$ and the lattice vectors are indicated in Figure (c).

be derived from the periodicity vector $\vec{V}$ and the two components of $\vec{V}$, as illustrated in Figure 3.3(c). An explanation of the phyllotactic notation is provided in Section 2.3.

In Figure 3.3(b) for the corresponding contact network, each blue or red dot on the surface of the cylinder represents a sphere, and each black line corresponds to a contact between adjacent spheres. Every sphere is in contact with six other spheres.

Rolling out this contact network into a plane of vertical position z and angular position θ results in a regular hexagonal (or triangular) lattice [Figure 3.3(c)]. Each dot in this pattern represents a sphere, and each line corresponds to a contact between adjacent spheres.

A regular hexagonal lattice such as the one shown in Figure 3.3(c) is therefore characteristic for a uniform structure. Such a structure corresponds to a maximisation of the number of contacts. For any uniform structure (l, m, n) in general, the phyllotactic pattern is only a rotation of the hexagonal lattice. Each uniform structure is distinguished by its periodicity vector $\vec{V}$, which is determined by the phyllotactic triplet (l, m, n) (Section 2.3).

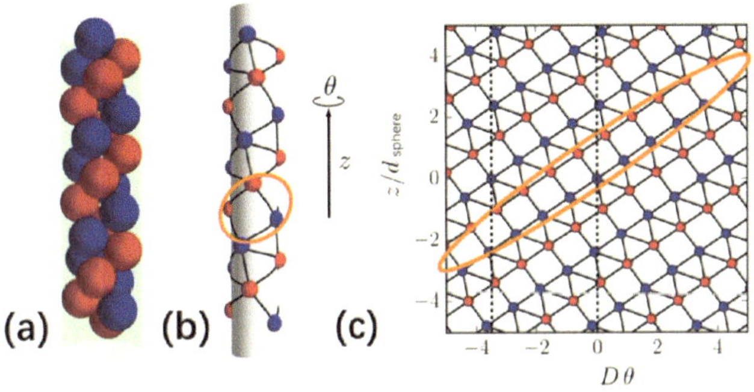

Figure 3.4 Illustration of a line-slip structure. Figure (a) shows the three-dimensional packing of a (3, **2**, 1) line slip. Figure (b) displays the corresponding contact network, where each dot represents a sphere and each line corresponds to a contact between adjacent spheres. Highlighted here is the gap or loss of contact between certain adjacent spheres. Figure (c) displays the contact network as rolled-out into a plane of height z and azimuthal angle θ. It demonstrates that the gaps or loss of contacts occur along a line between the red and the blue spheres, hence the name *line slip*.

In the hard-sphere limit, each uniform structure can only arise at a singular value of the diameter ratio D. For specific uniform structures with a given number of spheres lying on the same vertical position, the corresponding values of D can be derived analytically from the two-dimensional packing of a given number of boundary-touching circles inside a larger circle [Mughal *et al.* (2012)].

3.3.2 What Is a *Line-Slip* Structure?

For each uniform structure, there exists a related but different type of structure. We refer to each of such structures as a line slip. Its features are illustrated in Figure 3.4. Possible applications of such structures are discussed in Subsection 2.4.3.

The differences between a uniform and a line-slip structure are marginal, and difficult to spot from the corresponding three-dimensional packings [compare Figure 3.4(a) with Figure 3.3(a)]. However, a comparison between Figure 3.4(b) and 3.3(b) reveals a difference in contact network between a line slip and a uniform structure: Certain lines are missing, such as the one highlighted in Figure 3.4(b). Each sphere in Figure 3.4(b) is again represented by a red or a blue dot, and each line represents a contact between adjacent spheres. While every sphere in a uniform structure is in contact with six other spheres, the spheres in a line slip do not share the same number of contacts with adjacent spheres. For the $(3, \mathbf{2}, 1)$ line slip, some spheres have five contacts and others have six.

Rolling out the contact network in the z–θ plane yields the phyllotactic pattern in Figure 3.4(c). A comparison of this phyllotactic pattern with the one in Figure 3.3(c) reveals that such losses of contacts occur along a line; hence the name *line slip*. This feature was first identified by Picket *et al.* [Pickett *et al.* (2000)], but the term *line slip* was coined a decade later by Mughal *et al.* [Mughal *et al.* (2011, 2012)].

The bold number in the phyllotactic notation of a line slip indicates the direction in which the loss of contacts occurs. For a $(3, \mathbf{2}, 1)$ line slip [Figure 3.4(c)], the loss of contacts occurs along the **2** direction. Shearing continuously the row of red dots against its adjacent row of blue dots will end up with one of the two possible uniform structures. Thus we can say that each line slip is related to two adjacent uniform structures, one at a larger and one at a smaller value of D. Here, we adopt the convention of labelling each line slip by the related uniform structure of the *smaller* value of D [Mughal *et al.* (2012)].

Such shearing in the hard-sphere limit is accompanied by a continuous variation in the diameter ratio D. This explains why line-slip structures are intermediate between their uniform counterparts. In contrast to the cases of uniform structures at singular values of D, a line-slip structure in the hard-sphere limit exists across a finite range of D values (Table A.1 in Appendix A). As described in detail in Section 3.4, a line slip always has a lower packing fraction than its related uniform structures.

Since any line slip exists across a finite range of D, the gap between adjacent spheres in a line slip can be fine-tuned via a variation of the diameter ratio D or other structure-dependent parameters (such as the pressure of the system). This gives line slips special features with a huge potential for applications. Physical quantities such as packing fraction, stiffness, or conductivity can be adjusted by varying the gap size of the line slip.

Wood investigated the bending stiffness of cylindrical crystals in his thesis [Wood (2017)]. He employed molecular dynamics (MD) simulations to assemble spheres onto the surface of a cylindrical rod, and then calculated the stiffness of the rod after bending it into a variety of curvatures.

Electromagnetic properties such as electrical conductivity or resonance frequency may be calculated theoretically for line-slip structures using Kirchhoff's laws. Based on an electrical-circuit model [Ma (2020); Winkelmann (2020); Ma and Chan (2021)] that assumes each contact to be a resistor and each sphere to be a knot that connects adjacent resistors, the electrical conductivity of a line-slip structure can be studied as a function of the line-slip's gap size and of the diameter ratio D. On the other hand, one can also consider a different electrical-circuit model [Fan *et al.* (2010); Manoharan (2019)] where each sphere in a line-slip structure is assumed to be an inductor and each inter-sphere gap to be a capacitor. In this context, a line-slip structure represents a complex-network version of a LC circuit, where the resonance frequency of the structure can also be studied using Kirchhoff's laws.

Such special features of a line slip offer a huge potential for the development of novel materials such as microrods or polymers [Subsection 2.4.3]. Stiffness and conductivity play an important role in those materials, especially for microrods that make up liquid crystals, and both properties can be optimized via a manipulation of the gap sizes of line-slip structures.

3.4 Densest Hard-Sphere Packings inside Cylinders

Mughal *et al.* studied the densest columnar structures of hard spheres via the computational method of simulated annealing (Subsection 3.5.2) [Mughal *et al.* (2011, 2012); Mughal (2013)]. The structures obtained are all screw-periodic, and can be described by the corresponding unit cells. Such structures have been studied in detail up to a diameter ratio of $D \equiv D_{\text{cylinder}}/d_{\text{sphere}} = 2.873$ [Mughal *et al.* (2012)]. Table A.1 of Appendix A includes a list of all discovered structures. It also includes some structures with internal spheres that are not in contact with the cylindrical wall.

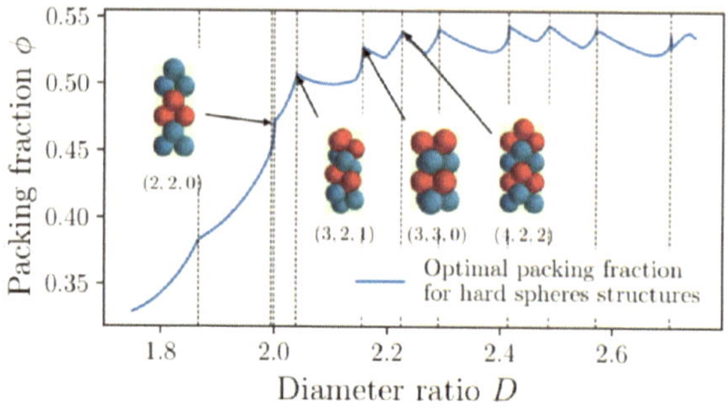

Figure 3.5 Packing fraction ϕ for hard spheres as a function of the diameter ratio $D \equiv D_{\text{cylinder}}/d_{\text{sphere}}$ for the densest packings of hard spheres [Mughal *et al.* (2012)]. The local maxima (black arrows) correspond to the uniform structures, and the curves in between correspond to the intermediate line-slip structures. The subjects of line-slip structures and uniform structures are discussed in Section 3.3.

The packing fractions of these tabulated structures have been calculated as a function of D (Figure 3.5). At local peaks of the packing fraction (dashed vertical lines) are the *uniform* structures. In between those structures with locally maximum packing fractions are the intermediate line slips, each of which is less dense than their related uniform structures. As a result of geometric confinement, the packing fractions of these structure are all far below the value of $\pi/\sqrt{18} \approx 0.74048$ for bulk fcc or hcp crystals.

Mughal *et al.* [Mughal *et al.* (2012)] also discovered that structures where all spheres are in contact with the cylinder wall are connected to disk packings on the unrolled surface of a cylinder. For $D > 2$, the disk packings then resemble the same contact network as their three-dimensional counterparts. Such connection can be understood by looking at a cut of the cylinder through the centre of a sphere. At small values of D, this cut is elliptical in shape. For increasing D, the curvature of the cylinder diminishes such that the cut approaches the shape of a circular disk. A detailed mathematical description of this transition is presented in Sections 4.1.1 and 4.1.2.

The reduction of this three-dimensional problem of sphere packing to a two-dimensional problem of disk packing allows one to

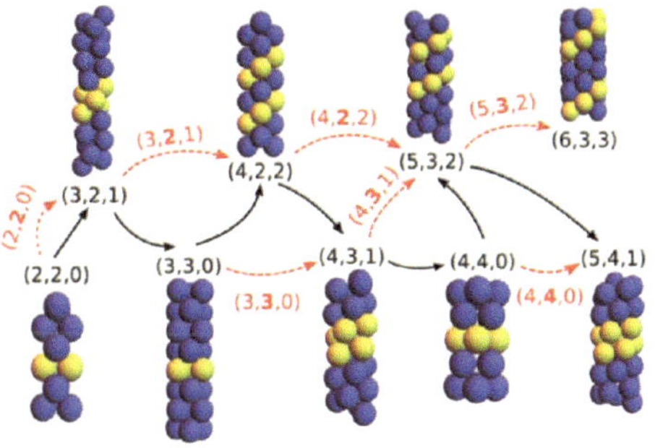

Figure 3.6 Sequences of simulated structural transitions between different uniform structures of hard spheres [Fu *et al.* (2017)]. The black lines indicate the sequence of densest-packed uniform structures for increasing diameter ratio *D*. Each red dotted arrow represents a dynamically favourable transition, as observed in simulations. Figure reprinted, with permission, from Ref. [Fu *et al.* (2017)].

define rules for possible transitions between uniform structures of different phyllotactic notations. Each uniform structure is connected to some other uniform structures through line slips. All possible transitions are summarised in a diagram for the three-dimensional sphere-packing problem and in another diagram for the two-dimensional disk-packing problem [Mughal *et al.* (2012)]. Since we refer to these two diagrams frequently, they are included in Appendix A of this monograph. They display the surface density (in the two-dimensional description) and the packing fraction (in the three-dimensional description) for all the uniform structures as well as their stable and also metastable intervening line slips.

Fu *et al.* [Fu *et al.* (2016)] vastly extended Mughal *et al.*'s table of densest hard-sphere packings up to $D = 4$ via a linear programming algorithm (see Subsection 3.5.2). In this extended regime of *D*, a total of seventeen new dense structures with internal spheres that are not in contact with the cylindrical wall have been discovered. Fu *et al.* [Fu *et al.* (2017)] also investigated the structural transitions between different uniform structures via the same method of linear programming, with their results summarised in a schematic plot (Figure 3.6).

Each image of a columnar structure in Figure 3.6 corresponds to a unique uniform structure as labelled by a corresponding phyllotactic notation (l, m, n). Each black arrow indicates the sequential appearance of uniform structures for increasing D according to the sequence of densest-packed structures (Figure 3.5). On the other hand, each red dashed arrow indicates the favourable sequence of structures for increasing D from a given uniform structure. In general, the sequence of structures as indicated by the red arrows is different from the sequence indicated by the black arrows. For instance, the $(3, 2, 1)$ uniform structure favours a direct transition to the $(4, 2, 2)$ uniform structure through the $(3, \mathbf{2}, 1)$ line slip for increasing D. However, Figure 3.5 shows that, in the sequence of densest packings, the $(3, 2, 1)$ uniform structure undergoes a transition to the $(3, 3, 0)$ uniform structure first before proceeding to the $(4, 2, 2)$ uniform structure. These investigations from cases of increasing D to cases of decreasing D have been extended by Mughal *et al.* [Mughal *et al.* (2018)]. We present in 5.4.3 a complete theoretical description of such transitions.

3.5 Simulation Techniques: Minimisation Algorithms

A variety of computational algorithms have been employed to simulate columnar structures. These include search-tree construction [Stoyan and Yaskov (2010)], molecular dynamics (MD) simulations [Lee *et al.* (2017); Wu *et al.* (2017)], as well as minimisation or optimisation algorithms such as simulated annealing [Mughal *et al.* (2012)], linear programming [Fu *et al.* (2016)], and sequential deposition [Chan (2011, 2013); Chan *et al.* (2019)]. In this section, we introduce a few important examples of minimisation or optimisation algorithms. In these computational approaches, columnar structures are obtained by either minimising a potential (such as the energy or enthalpy) or searching for the densest hard-sphere packing, starting from a given initial configuration.

Optimisation algorithms can generally be classified into two categories: *Local* routines and *global* routines. A local routine involves a search for the nearest optimum from a given initial position or configuration, and a global routine involves a search for the optimum throughout the whole parameter space. Each optimisation problem can always be interpreted as a minimisation problem. In this section, several different algorithms are discussed in terms of the minimisation of a particular multidimensional function $E(\vec{X})$. This function, which contains information about the spheres' coordinates, can be the energy, enthalpy, or unit cell length of the simulated system.

3.5.1 Local Minimisation Routines

Gradient descent: The simplest algorithm in finding the (local) minimum of a function $E(\vec{X})$ is the gradient-descent (or steepest descent) algorithm. It is based on the gradient $\vec{\nabla}E(\vec{X})$ of this function (see Appendix B for a definition of the gradient).[1] The gradient taken at a given parameter vector $\vec{X}$ corresponds to the direction of steepest ascent. By iteratively taking steps into the negative direction of the gradient, one eventually ends up in a local minimum of the function where the gradient is zero.

The strength of the gradient-descent method lies in its simplicity and robustness. It can easily be implemented and not much information about the multidimensional function $E(\vec{X})$ is necessary. The only additional information required is the gradient of $E(\vec{X})$, which can be calculated numerically. However, as a drawback of the algorithm, it would usually take many computational steps for the gradient to converge to zero in the vicinity of the minimum.

BFGS algorithm: A more sophisticated local minimisation routine is the BFGS method, named after Broyden, Fletcher, Goldfarb, and Shanno [Byrd *et al.* (1995)]. It belongs to the class of quasi-Newton algorithms. It employs the Hessian matrix of the function $E(\vec{X})$ as additional information to locate the nearest minimum. The Hessian contains information about the second derivative of the function to be minimised (see Appendix B for the definition of the Hessian).

[1] When the function to be minimised is a potential, the gradient can be interpreted as a force vector of the system.

A Newton method that approximates the Hessian numerically is referred to as a quasi-Newton method.

The minimum is found in an iterative procedure, in which a matrix equation involving the Hessian is solved. The BFGS method directly approximates the inverse of the Hessian and terminates the iteration when the gradient becomes zero.

The BFGS method is much more efficient in finding the nearest local minimum than the gradient-descent method. Its efficiency comes at a cost of complexity, however. Additional information such as the Hessian of the function $E(\vec{X})$ is required. This method is explained in detail in Appendix B, where each step of the algorithm is explained with a pseudo-code.

3.5.2 Global Minimisation Routines

Linear programming: Fu *et al.* have obtained results of columnar hard-sphere packings via an optimisation routine called linear programming [Fu *et al.* (2016); Fu (2017); Fu *et al.* (2017)]. They have managed to significantly extend Mughal *et al.*'s investigations on densest structures up to $D = 4$. This is discussed in detail in Section 3.4.

The method of linear programming is used for optimisation problems with the following goal: *Minimise a linear objective function subject to linear equality and inequality constraints.*

For packings of hard spheres in cylindrical confinement, the function to be minimised is the volume of the unit cell, which depends on its length L. Searching for the corresponding densest packings can be formulated as the following problem of linear programming: Minimise L subject to the following pair of constraints: (1) No overlap between spheres, and (2) no overlap of any sphere with the cylindrical wall. The method of linear programming is fast and efficient, but it is also a challenge to locate the global minimum of a function with a high accuracy. Due to the constraint of zero inter-sphere overlap, this method is restricted to simulations of hard-sphere packings.

Simulated annealing: Mughal *et al.* [Mughal *et al.* (2012)] employed the method of simulated annealing to search for the densest packings of hard spheres (Section 3.4). This is a probabilistic routine that seeks the global minimum of a given function. Its approach is based on the metallurgic process of heating a metal and cooling it slowly down to avoid defects or metastable states [Metropolis *et al.* (1953)].

The spheres, which are initially placed at some random positions, undergo random trial movements in each computational step. If the movements lead to a decrease in the value of the function, they would be accepted. Otherwise, they would be accepted only at some temperature-dependent probability criterion. The most common acceptance criterion is the Metropolis criterion (Appendix B), which was also used in Mughal *et al.*'s investigations. During an optimisation process, the temperature is gradually decreased according to an annealing schedule. In Mughal *et al.*'s simulations, this schedule was either linear (i.e., decreasing the temperature at fixed intervals) or logarithmic (e.g., $0.1, 0.09, 0.08 ... 0.01, 0.009, 0.008 ...$). At $T = 0$, the system is supposed to have settled at the global minimum of the function.

With a sufficiently slow annealing schedule, the correct global minimum can be attained accurately. Similar to the gradient-descent algorithm, its greatest advantages are its simplicity and robustness. However, at zero temperature when only movements that lower the function's value are accepted, the method of simulated annealing becomes equivalent to the gradient-descent algorithm, with the same disadvantages of being slow and inefficient.

This algorithm was employed to minimise the energy of an initial packing of soft spheres [Mughal *et al.* (2012)]. A hard-sphere packing is generated when this soft-sphere energy becomes zero without any overlapping between spheres. After repeated runs, the structure with the highest packing fraction is then chosen to create Figure 3.5.

Basin-hopping: The method of simulated annealing can be improved via the combination with a local-minimisation method. This is the basic idea of the basin-hopping method [Wales and Doye (1997); Jones *et al.* (2001)]. Its search for the global minimum follows a similar probabilistic approach, but during each iteration an additional process of local minimisation is involved.

Each iteration consists of a cycle with three important steps:

(1) Random perturbation of the parameter vector $\vec{X}$
(2) Performance of local minimisation
(3) Accept or reject the new parameter vector based on the new local minimum.

The acceptance criterion is similar to that of simulated annealing. The new parameter vector is accepted if the new minimum is lower than its predecessor. Otherwise a probability criterion such as the Metropolis criterion is employed.

Since the basin-hopping algorithm involves a use of the BFGS method for local minimisation, it shares with the BFGS method a common advantage of arriving at the global minimum quickly and accurately. We employed this minimisation method wherever we are interested in stable structures with a global minimum in energy (or enthalpy). A detailed description of this method is provided in Appendix B.

Sequential deposition: A wide range of densest possible columnar structures of hard spheres can be obtained computationally or experimentally through a method of sequential deposition [Chan (2011, 2013); Chan *et al.* (2019)]. Consider the construction of a columnar structure within a vertically aligned cylindrical tube. By dropping the spheres one by one onto their lowest possible positions, we would achieve a local maximum of the packing fraction everywhere across the system. This packing fraction can be fine-tuned into a global maximum by varying the spatial configuration of the first few spheres at the bottom, where we refer to this configuration as the underlying template of the structure. A detailed description of this method is provided in Chapter 4.

Chapter 4

Packing of Hard Spheres by Sequential Deposition

Relevant journal publications by H.-K. Chan:

- H.-K. Chan, Phys. Rev. E 84, 050302(R) (2011).
- H.-K. Chan, Philos. Mag. 93, 4057-4069 (2013).
- H.-K. Chan, Y. Wang and H. Han, AIP Adv. 9, 125118 (2019).

This chapter describes how any densest possible columnar structure of hard spheres at diameter ratios $D \in [1,2.7013]$ can be obtained computationally or experimentally via a method of sequential deposition [Chan (2011, 2013)]. The concept of this method is to generate, for any given value of D, dense packings of identical spheres with all spheres being in contact with the inner surface of the confining cylinder. With every sphere being the same radial distance away from the central axis, the corresponding simulations benefit from a reduction of dimensionality from three to two. A two-dimensional lattice that takes into account the angular and axial (vertical) dimensions of the cylindrical polar coordinate system is sufficient for modeling such a packing process in three-dimensional space. This turns the packing algorithm into a computationally efficient

Columnar Structures of Spheres: Fundamentals and Applications
Jens Winkelmann and Ho-Kei Chan

ISBN 978-981-4669-48-1 (Hardcover), 978-0-429-09211-4 (eBook)
www.jennystanford.com

one. The chapter begins with an explanation of our mathematical considerations, before proceeding to a description of the corresponding sphere-packing algorithm. [1]

4.1 Mathematical Description of Sequential Deposition

In the first section of this chapter we provide the mathematical details that are required for understanding the algorithm of sequential deposition of hard spheres. It begins with a geometric description of two identical spheres in contact. This is followed by a derivation of the mathematical representation that describes the projection of each cylinder-touching sphere on the unrolled surface of the cylinder. We then describe mathematically the condition for the diameter ratio D at which multiple spheres can be deposited simultaneously at the same vertical position. We end the section by defining empirical packing fractions for an infinitely long cylinder. These packing fractions apply regardless of whether the corresponding columnar structures are periodic. They work for the specific cases of periodic structures in which all spheres are in contact with the cylindrical surface.

4.1.1 Equations for the Geometric Relation between Two Spheres in Contact

For this chapter, consider expressing all length quantities in units of the sphere diameter. In this approach, the dimensionless sphere diameter is equal to unity, the dimensionless cylinder diameter is equal to the cylinder-to-sphere diameter ratio D, the dimensionless radial position of each cylinder-touching sphere is equal to $(D - 1)/2$, and the dimensionless centre-to-centre separation between two spheres in contact is equal to a dimensionless sphere diameter, i.e., unity. The dimensionless cylinder diameter D, and the dimensionless radial position $(D - 1)/2$ of each cylinder-touching

[1] A detailed description of the method of simulation is provided in Ref. [Winkelmann and Chan (2020)].

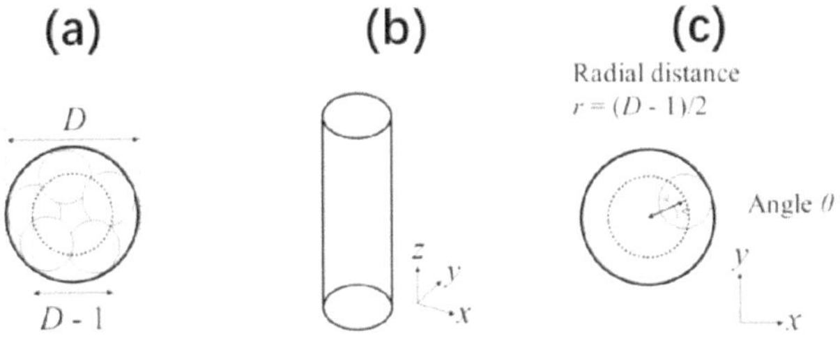

Figure 4.1 Schematic illustration of various geometric parameters: (a) Cross-sectional view showing the dimensionless cylinder diameter D, and the diameter $D - 1$ of an inner cylindrical surface that embeds all the sphere centres; (b) Cartesian coordinates (x,y,z) with the z direction lying along the axis of the cylinder; (c) Cross-sectional view showing the position of a sphere's centre in polar coordinates, where the sphere's centre is at an angular position θ and at a radial position $r = (D - 1)/2$.

sphere, are respectively illustrated in Figure 4.1(a) and Figure 4.1(c), both for a cross-sectional view of the cylindrical tube.

For $i = \{1, 2\}$, let (x_i, y_i, z_i) be the Cartesian coordinates of the centres of an arbitrary pair of spheres, where z_i lies along the axis of the cylinder [Figure 4.1(b)]. The general condition for them to be in contact is given by

$$(x_1 - x_2)^2 + (y_1 - y_2)^2 + (z_1 - z_2)^2 = 1 \tag{4.1}$$

Since both spheres are in contact with the surface of the cylinder, the Cartesian coordinates x_i and y_i are related to the angular position θ_i in the cylindrical polar coordinate system by

$$x_i = \left(\frac{D-1}{2}\right)\cos\theta_i \quad \text{and} \quad y_i = \left(\frac{D-1}{2}\right)\sin\theta_i, \tag{4.2}$$

respectively, as illustrated in Figure 4.1(c). By substituting Equation (4.2) into Equation (4.1) and applying various trigonometric identities, we obtain

$$(\Delta z)^2 + (D-1)^2 \sin^2\left(\frac{\Delta\theta}{2}\right) = 1 \tag{4.3}$$

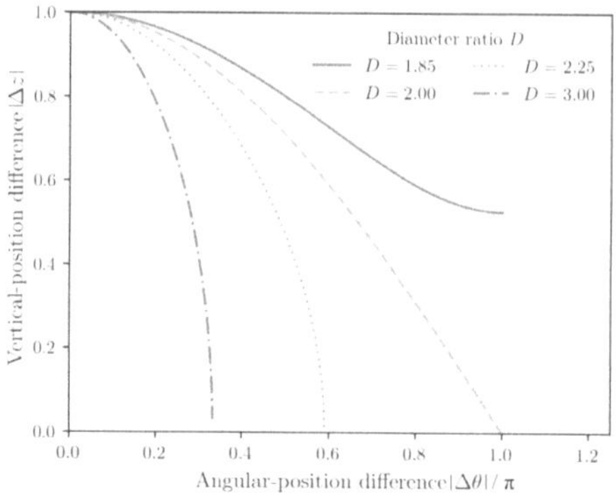

Figure 4.2 Plot of $|\Delta z|$ against $|\Delta\theta|/\pi$ for various values of the diameter ratio D based on Equation (4.5). As illustrated in the figure, a solution for $|\Delta z| = 0$ exists only at $D \geq 2$, meaning that at $D < 2$ two spheres cannot be placed at the same z position.

where

$$\Delta z \equiv z_2 - z_1 \quad \text{and} \quad \Delta\theta \equiv \theta_2 - \theta_1. \tag{4.4}$$

Equation (4.3) describes two spheres in contact, where both spheres are also in contact with the inner surface of the cylinder. For $0 \leq |\Delta\theta| \leq \pi$, it follows that

$$|\Delta z| = \sqrt{1 - (D-1)^2 \underbrace{\left(\frac{1 - \cos|\Delta\theta|}{2}\right)}_{\sin^2\left(\frac{|\Delta\theta|}{2}\right)}} \tag{4.5}$$

From Figure 4.2 where Equation (4.5) is plotted, it can be seen that, at $D < 2$, there is no solution for $|\Delta z| = 0$, meaning that any pair of spheres cannot be placed at the same z position. Mathematically, the condition of $|\Delta z| = 0$ corresponds to

$$|\Delta\theta| = \left|\cos^{-1}\left[1 - \frac{2}{(D-1)^2}\right]\right|, \tag{4.6}$$

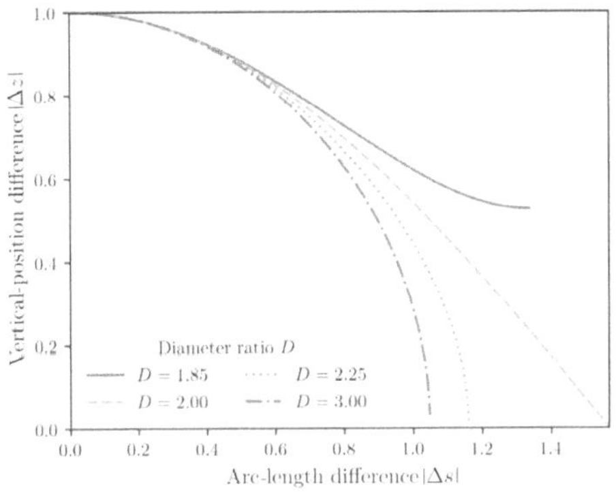

Figure 4.3 Plot of $|\Delta z|$ against $|\Delta s|$ for various values of the diameter ratio D based on Equation (4.12). As the value of D increases, the shape of the curve becomes more circular-like, because Equation (4.12) approaches the limiting Equation (4.14) for a circle. For any case of $D \leq 2$, the maximum value of $|\Delta s|$ corresponds to the upper limit $|\Delta\theta| = \pi$ (compare with Figure 4.2).

which implies

$$-1 \leq 1 - \frac{2}{(D-1)^2} \tag{4.7}$$

$$\text{and } 1 - \frac{2}{(D-1)^2} \leq 1. \tag{4.8}$$

It can be shown from inequality (4.7) that only at

$$D \geq D_{\text{min}} = 2 \tag{4.9}$$

can any pair of spheres be placed at the same z position. The inequality (4.8) implies

$$\frac{2}{(D-1)^2} \geq 0 \tag{4.10}$$

which is valid for any value of $D > 1$. It follows that there does not exist a corresponding upper limit for D.

Consider an arc length

$$s \equiv \left(\frac{D-1}{2}\right)\theta \tag{4.11}$$

which corresponds to positions on the cylindrical surface [Figure 4.1(a)] that embeds all the sphere centres. Equation (4.3) can then be written as

$$(\Delta z)^2 + \frac{(D-1)^2}{2}\left[1 - \cos\left(\frac{2\Delta s}{D-1}\right)\right] = 1 \tag{4.12}$$

where

$$\Delta s \equiv s_2 - s_1. \tag{4.13}$$

According to the Taylor expansion of $\cos(x)$ at $x \ll 1$, Equation (4.12) approaches

$$(\Delta z)^2 + (\Delta s)^2 = 1 \tag{4.14}$$

at $D \to \infty$, i.e., at $D - 1 \gg 2\Delta s$. Equation (4.14) is just the equation of a circle and represents a disappearance of all geometric influences from the cylinder, i.e., a case of packing circular disks on a plane. The approach to this limit for increasing D is demonstrated in Figure 4.3.

4.1.2 Surface Representation of a Single Sphere

Consider, again, the inner cylindrical surface, of diameter $(D - 1)$, that embeds all the sphere centres [Figure 4.1(a)]. Since this curved surface contains all information about the positions of the spheres, one can imagine unrolling it into a flat surface and then analyse the closed-packed three-dimensional structure of the spheres. At the planar limit $D \to \infty$, each sphere occupies a circular region of radius 1/2 (in units of the spheres' diameter), and the circular boundary of such a region is described by the equation

$$z'^2 + s'^2 = \left(\frac{1}{2}\right)^2 \tag{4.15}$$

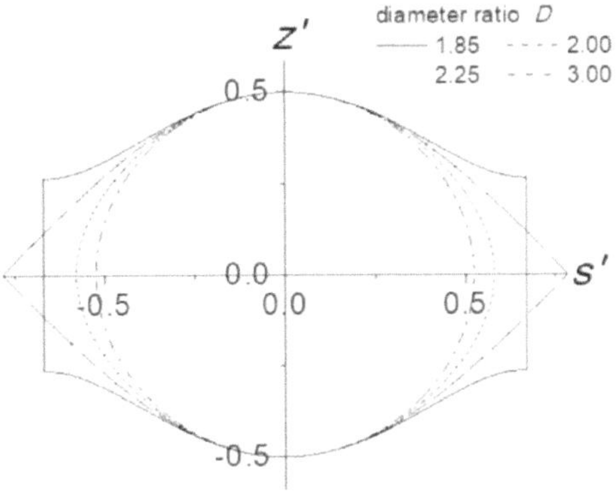

Figure 4.4 Plot of z' against s' for various values of the diameter ratio D. The axes z' and s' are the Cartesian coordinates in the unrolled cylindrical surface with respect to the centre of the circular region. As the value of D increases, the region that represents a sphere on the unrolled cylindrical surface approaches the circular limit as described by Equation (4.15). For $D < 2$, each "gap" that arises from the lack of a solution at $z' = 0$ is bridged by a vertical line, forming a closed region that is defined to represent the sphere on the unrolled surface.

where z' and s' are Cartesian coordinates with respect to the centre of the circular region. This equation can be rewritten as

$$(2z')^2 + (2s')^2 = 1 \tag{4.16}$$

which corresponds to an application of the transformations

$$\Delta z \to 2z' \quad \text{and} \quad \Delta s \to 2s' \tag{4.17}$$

to Equation (4.14).

Let the region occupied by each sphere on this unrolled surface be generally defined by these transformations. The boundary of such a region can then be obtained by applying Equation (4.17) to Equation (4.12), which results in

$$(2z')^2 + \frac{(D-1)^2}{2}\left[1 - \cos\left(\frac{4s'}{D-1}\right)\right] = 1. \tag{4.18}$$

This implies

$$z' = \pm\frac{1}{2}\sqrt{1 - \frac{(D-1)^2}{2}\left[1 - \cos\left(\frac{4s'}{D-1}\right)\right]} \quad (4.19)$$

Similar to Equation (4.5), no solution exists for $z' = 0$ at $D < 2$. For such cases, each corresponding 'gap' in a graphical representation of Equation (4.19) is bridged by a vertical line as shown in Figure 4.4, forming a closed region that represents the sphere on the unrolled cylindrical surface. Such a region approaches the circular limit for increasing D, and at $D = 3$, it already mimics a disk that appears to be slightly stretched along the s' axis. Note that the surface representations of two touching spheres also appear to be in mutual contact on this unrolled surface.

4.1.3 Simultaneous Deposition of Spheres

Consider a simultaneous deposition of u spheres into regular angular spacings and into the same z position. The angular separation between any two neighbouring spheres is given by

$$(\Delta\theta)_u \equiv \frac{2\pi}{u}, \quad (4.20)$$

provided that D is greater than a threshold $D_{\text{min},u}$ such that the set of u spheres can all be placed at the same z position. At $D = D_{\text{min},u}$, any pair of neighbouring spheres are in mutual contact so that the relation

$$\cos(\Delta\theta)_u = 1 - \frac{2}{(D_{\text{min},u} - 1)^2} \quad (4.21)$$

holds if it is $u \geq 2$, according to Equation (4.6) for the condition of $\Delta z = 0$. The value of $D_{\text{min},u}$ for $u \geq 2$ is then given by

$$D_{\text{min},u} = 1 + \sqrt{\frac{2}{1 - \cos(2\pi/u)}} \quad (4.22)$$

which is consistent with Equation (4.9) for $u = 2$.

In fact, the definition of $D_{\min,u}$ as the minimum value of D for which u spheres can be placed at the same z position can be extended to include the trivial case of $D_{\min,u} = 1$ for $u = 1$. In general, for $u \geq 1$, the simulation only needs to take into account an angular range of

$$0 \leq \theta \leq \frac{2\pi}{u}, \tag{4.23}$$

or equivalently

$$0 \leq s \leq \frac{(D-1)\pi}{u}, \tag{4.24}$$

because the structure repeats itself along the angular dimension.

4.1.4 Empirical Packing Fractions for an Infinitely Long Cylinder

For a finite-sized system, the volume occupied by the spheres and the total volume of the container are both finite, so that the concept of a packing fraction is well-defined, as described in Subsection 3.1. However, for the present packing problem, the container is ideally an infinitely long cylindrical tube, and therefore the packing fraction would be ill-defined if we stick to the definition mentioned above.

There are, however, alternative ways of defining a packing fraction for an infinitely long cylindrical container. For example, one can consider (i) a finite section of the cylinder, where the close-packed structure of spheres is a repeating unit [Mughal *et al.* (2011, 2012)] along the axis of the cylinder, or (ii) a sufficiently long section [Chan (2011)] of the cylinder, where "end effects" would not make any practical difference to the evaluated packing fraction. Since there is no way to know *a priori* whether a close-packed structure of spheres inside an infinitely long cylindrical tube is periodic or not, the latter empirical definition of packing fraction for a sufficiently long column of spheres is more appropriate for a general applicability. According to Equation (3.1), the packing fraction defined in this manner can be expressed as

$$\phi \equiv \frac{2}{3D^2} \lim_{L\to\infty} \left(\frac{N}{L}\right) \tag{4.25}$$

where the length L is expressed in units of the spheres' diameter d_{sphere}. The quantity (N/L) is the number of spheres per unit length along the axis of the cylinder. Its value can be evaluated from a plot of N against z, where N is the number of spheres as counted from one end of the tube section, and z is the position of a sphere along the axis of the cylinder. The slope of such a plot as obtained from a linear fit would be a good approximate value for (N/L).

On the other hand, since all spheres are in contact with the cylindrical boundary, one can also consider an alternative packing fraction, defined as the volume of spheres divided by the cylindrical ring that contains the spheres, where such a cylindrical ring exists only at $D > 2$ [Figure 4.1(a)]. This packing fraction is given by

$$\phi_{\text{ring}} \equiv \lim_{L\to\infty} \frac{N\left[\frac{4}{3}\pi\left(\frac{1}{2}\right)^3\right]}{L\left[\pi\left(\frac{D}{2}\right)^2 - \pi\left(\frac{D-2}{2}\right)^2\right]} = \frac{1}{6(D-1)} \lim_{L\to\infty}\left(\frac{N}{L}\right). \quad (4.26)$$

While the value of ϕ would approach zero at the planar limit of $D \to \infty$, the value of ϕ_{ring} would remain finite at any arbitrarily large value of D, making itself a generally useful measure of packing fraction for the present surface-packing algorithm where all spheres are in contact with the cylindrical boundary. For the simultaneous deposition of spheres described above, Equation (4.25) and Equation (4.26) should respectively be written as

$$\phi = \frac{2}{3D^2} \lim_{L\to\infty}\left[u\left(\frac{N}{L}\right)_u\right] \quad (4.27)$$

and

$$\phi_{\text{ring}} = \frac{1}{6(D-1)} \lim_{L\to\infty}\left[u\left(\frac{N}{L}\right)_u\right], \quad (4.28)$$

where $(N/L)_u$ is the number of spheres per unit length in the corresponding simulation array for a repeating pattern of the structure.

4.2 Empirical Trials of Sequential Deposition

Equation (4.25) and Equation (4.26), which relate the corresponding packing fractions to the number of spheres per unit length along the z axis, suggest how densest possible packings of equal-sized spheres inside a cylindrical tube may be constructed. They imply that, for any given value of D, a maximisation of the corresponding packing fraction is equivalent to a maximisation of the number of spheres per unit length, which means that one only needs to be concerned with the number density along the vertical z axis.

A possible approach is therefore to deposit spheres, one by one, into positions of smallest possible z, where this approach ensures that in each deposition step the local structure involving the addition of a sphere is densest possible. Since the whole column of spheres is made up of locally densest regions, it is not unreasonable to suppose that the whole column of spheres is also at its densest possible configuration. This is formally described by the following conjecture, which relates local packing processes to the global packing fraction: *If the packing of spheres is locally densest everywhere in the columnar structure, the overall structure would be at its globally densest configuration.*

4.2.1 Successful Trial for a Densest Zigzag Structure

Take the zigzag structure at $D = 1.3$ as an example. If a first sphere is placed at $z = 0$ and $\theta = 0$, the angular position of smallest z for the deposition of the second sphere is $\theta = \pi$, and for the third sphere it is back to $\theta = 0$, and so on. Such an alternation between $\theta = 0$ and $\theta = \pi$ constitutes the formation of a zigzag structure (Figure 4.5), with all the sphere centres lying on a plane as defined by these two angles. For this reason, the structure is also referred to as a planar zigzag. This structure is the densest possible configuration for $D = 1.3$, which has been corroborated computationally by simulated annealing [Pickett *et al.* (2000); Mughal *et al.* (2011, 2012)]. The present example shows that it is possible to construct a densest possible structure via a surprisingly simple procedure of sequential deposition.

4.2.2 Successful Trial for a Densest Achiral Structure

Consider the structure of achiral doublets at $D = 2$ (Figure 4.6). For a first sphere placed at $z = 0$ and $\theta = 0$, the position of smallest z for the second sphere corresponds to $z = 0$ and $\theta = \pi$, where $z = 0$ is defined to be the smallest possible value of z for any sphere, i.e., no sphere can be placed at $z < 0$. If the cylinder is held upright, this condition of $z \geq 0$ can practically be enforced via the use of a flat base that supports the spheres against gravity. Interestingly, for the third sphere, there exist two angular positions of smallest z, at $\theta = \pi/2$ and $\theta = 3\pi/2$, respectively. Such a problem can be resolved if we specify in our packing algorithm a rule to choose from multiple positions of smallest z, for example, by choosing the first of such positions encountered in a scan across all angular positions. The direction of such a scan may be defined as one of increasing or decreasing θ, and the initial angular position of the scan may be taken as that of

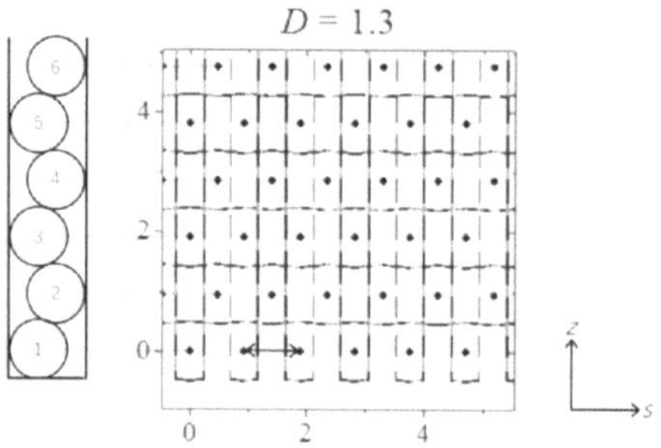

Figure 4.5 Schematic illustration of the process of sequential deposition for constructing the densest possible zigzag structure at $D = 1.3$. The numbers 1 to 6 indicate the sequence of the first six spheres in the deposition process. Each sphere is deposited into a position of smallest possible z, which is at an angular displacement of π from the position of the previously placed sphere. The corresponding surface representation of this structure is shown on the right, where the double arrow indicates a periodicity that corresponds to the cross-sectional circumference of the inner cylindrical surface. The dashed lines provide a one-to-one correspondence between the spheres in the schematic diagram and the points on the surface pattern that represent the sphere centres.

the previously placed sphere. Both the scan direction and the initial angular position are important because they would determine which position of smallest z would be first encountered in a scan.

For instance, if the scan is conducted all along the direction of increasing θ as indicated by the arrows in Figure 4.6, the third sphere should be placed at $\theta = 3\pi/2$, followed by a deposition of the fourth sphere at $\theta = \pi/2$. The deposition of the fifth to the eighth sphere follows the same procedure, except that the fifth sphere is to be placed at $\theta = \pi$, i.e., the same angular position as that of the second sphere. If this packing process continues indefinitely, the resulting structure (Figure 4.6) is one with each corresponding z position shared by a pair of spheres. This structure is described as a collection of achiral doublets, where each doublet of spheres is distinguished from its neighbouring doublets by a $\pi/2$ rotation. This example demonstrates the need of an additional rule for defining

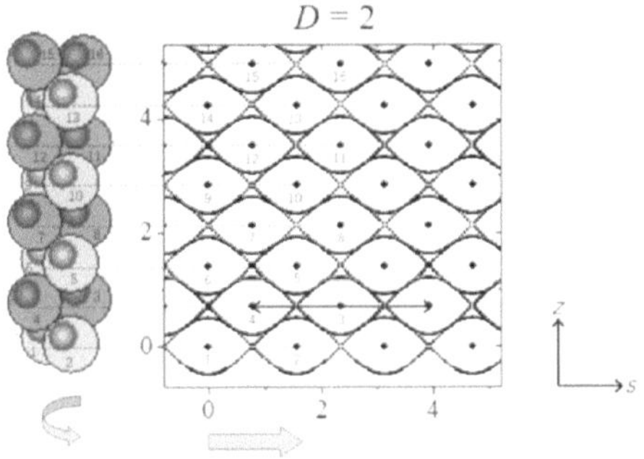

Figure 4.6 Schematic illustration of the process of sequential deposition for constructing the densest possible structure of achiral doublets at $D = 2$. The numbers 1 to 16 indicate the sequence of the first sixteen spheres in the deposition process. The final structure is one with each corresponding z position shared by a pair of spheres. The arrows indicate the direction of each angular scan for positions of smallest z, and the double arrow in the surface representation indicates a periodicity that corresponds to the cross-sectional circumference of the inner cylindrical surface. The dashed lines provide a one-to-one correspondence between the spheres in the three-dimensional illustration and the points on the surface pattern that represent the sphere centres.

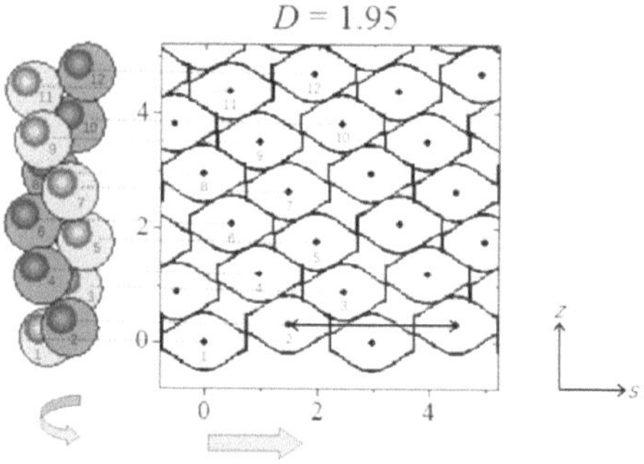

Figure 4.7 Schematic illustration of the process of sequential deposition for constructing a non-densest double-helix structure at $D = 1.95$, where the second sphere is placed in a lowest possible position at $\theta = \pi$. The numbers 1 to 12 indicate the sequence of the first twelve spheres in the deposition process. The resulting structure consists of two helices in different environments, which are respectively labelled by odd and even numbers.

our packing algorithm whenever multiple positions of smallest z are encountered.

4.2.3 Unsuccessful Trial for a Densest Single-Helix Structure

At this point, it appears that the computationally predicted densest possible structures can all be generated by the abovementioned approach of sequential deposition. There are exceptions, however. For example, at $D = 1.95$, it was found that this relatively simple approach does not yield the computationally predicted single-helix structure, but rather a slightly less dense double-helix structure (Figure 4.7) even though all the spheres are jam-packed inside the cylindrical tube. A possible reason for such a discrepancy is that the positional difference between the first two spheres does not conform to values of $\Delta\theta$ and Δz in the single-helix structure, assuming that these two spheres belong to the periodic structure. This suggests that, as a necessary condition, the values of $\Delta\theta$ and Δz for the first two spheres must coincide with the relative positions of any pair of spheres in the single helix for a generation of this densest possible structure. But more importantly, it also suggests that the resulting structure can be dependent, possibly in a sensitive manner, on the relative positions of the first two or more spheres deposited into the cylinder. This suggests a possible way of "fine-tuning" the resulting structure.

4.2.4 Successful Trial for a Densest Single-Helix Structure

Figure 4.8 illustrates how the densest possible single-helix structure at $D = 1.95$ can be recovered by changing the positional difference between the first two spheres, where these two spheres are in mutual contact as described by Equation (4.3). Figure 4.9 shows a plot of N against z for both structures where, as mentioned in the previous section, the average slope of such a plot provides a good approximate value for (N/L), i.e., the number of spheres per unit length along the cylinder axis. It can be seen that the slopes of the single- and double-helix structures are different, though not to a great extent. According to Equation (4.25), this reflects the slight difference in packing fraction between the two structures. In a recent paper by Ho-Kei Chan and his students [Chan *et al.* (2019)], an analytic theory is presented to explain how the densest possible helical structures at $D < 2$ can be obtained by optimising the positional difference between the first two spheres.

4.3 Problem with a Flat Base at $D > 2$

A slightly more complicated scenario exists for cases of $D > 2$: For each value of D, there exists a regime of θ where a second sphere can be placed at a position of smallest z at $z = 0$, i.e., it would not be supported by a first sphere below it. The spheres are either in mutual contact at the same z position or they cannot be brought into contact with each other at all, and the angular separation between the spheres is either equal to or greater than the critical value given by Equation (4.6). Such cases can be thought of as corresponding to some geometric influence from a flat base that supports the spheres inside a vertically aligned cylinder, as illustrated in the schematic diagrams of Figure 4.10. Since the flat base is not geometrically

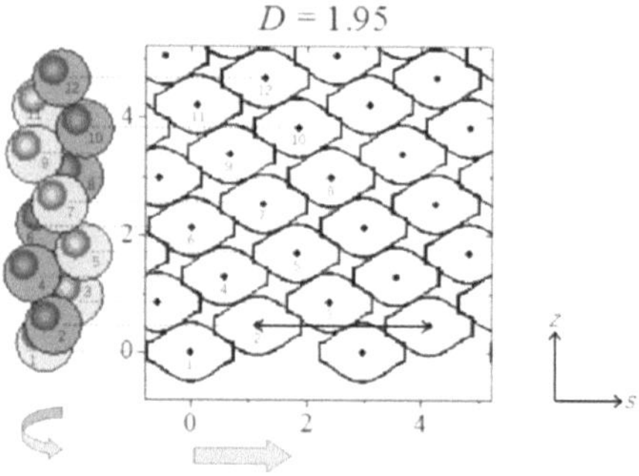

Figure 4.8 Schematic illustration of the process of sequential deposition for constructing the densest possible single-helix structure at $D = 1.95$, where the second sphere is not placed at $\theta = \pi$ but is placed at $\theta \approx 140.3°$. The numbers 1 to 12 indicate the sequence of the first twelve spheres in the deposition process. As different from the double helix, the environments of the two helices in this structure are the same, hence it is referred to as a single helix despite the presence of two helices in the structure.

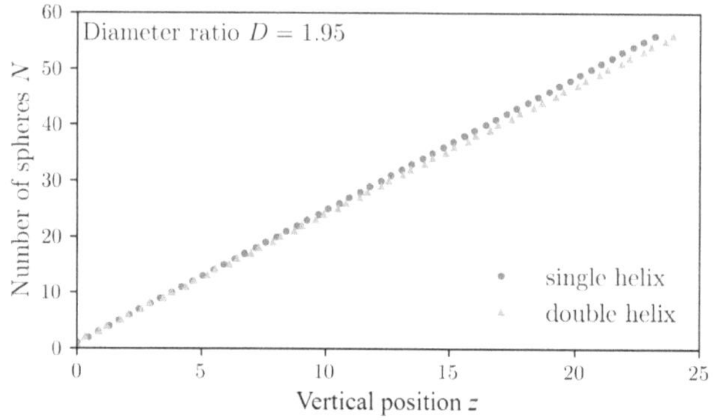

Figure 4.9 Plot of N against z for the double- and single-helix structures at $D = 1.95$, where N is the number of spheres as counted from one end of the tube section and z is the position of a sphere along the axis of the cylinder. As shown in the figure, the slope for the densest possible single-helix structure is greater than that for the double-helix structure, which reflects the slight difference in packing fraction between the two structures.

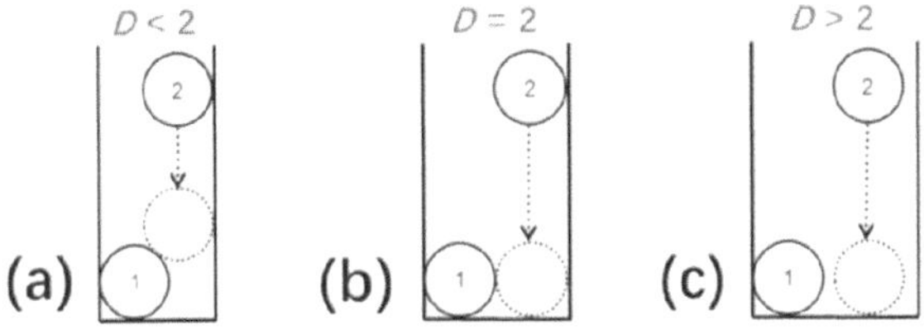

Figure 4.10 Schematic illustration of the role of a flat base for spheres inside a vertically aligned cylinder: (a) At $D < 2$, only the first sphere is in contact with the flat base, so that this flat base plays no role in the positions of subsequent spheres; (b) At $D = 2$, the second sphere, at its position of smallest z, is not only in contact with the flat base, but it is also in contact with the first sphere; (c) At $D > 2$, there exists a range of angular positions where the second sphere is in contact with the flat base (i.e., at $z = 0$) but not in contact with the first sphere. While in the case of $D = 2$ the flat base contributes to a formation of the densest possible structure of "achiral doublets" (Figure 4.6), for cases of $D > 2$ there is no guarantee that a deposition of spheres onto the flat base would result in a densest possible structure. This is because the flat base is not geometrically related to the desired densest possible structure.

related to the corresponding densest possible structure, there is no guarantee that the relative positions of the first few spheres, as influenced by the flat base, would coincide with the relative positions of spheres in the desired densest possible structure. It follows that there is a need to move the first few spheres around so as to "fine-tune" the resulting structure into the densest possible one, like the case of $D = 1.95$ described above. For such fine-tuning, if the second sphere is at an angular position such that it rests on the flat base, one should also vary the z position of the sphere so as to eliminate any geometric influence from the flat base on the sphere's position. If the value of D is large enough such that the third and subsequent spheres can also fall onto the flat base, additional rules are needed to specify how the positions of these spheres should be varied or how they should be determined from the position of the second sphere.

4.4 Sequential Deposition: The Packing Algorithm

A sphere-packing algorithm, which allows the positions of the third and subsequent spheres in a sufficiently wide cylindrical tube to be determined from the position of the second sphere, is presented here. In such an algorithm, the search for a densest possible structure would at most be a two-dimensional one, because only the angular position, and in certain cases also the z position, of the second sphere needs to be varied. This saves a lot of computational time, though at the expense of a thorough search space for the optimisation process.

4.4.1 Packing Algorithm for $D \in [1,2)$

For $D \in [1, 2)$, no sphere can come into contact with the flat base after a first sphere is placed at $z = 0$ and $\theta = 0$, according to Equation (4.9). It follows that only the angular position, but not the z position, of the second sphere needs to be varied. At any given angular position θ_2, the second sphere is to be placed at a position of smallest z such that it comes into contact with the first sphere. As illustrated in the examples of $D = 1.95$ (Figure 4.7 and Figure 4.8), the position of the second sphere can play a role in determining the resulting columnar structure of spheres if the third sphere is in contact with both the second and the first sphere. This is because, if the position of the second sphere is varied, the position of the third sphere, and hence also the positions of subsequent spheres, would be shifted accordingly. In principle, for any given value of D, the angular position θ_2 of the second sphere should be varied from 0 all the way to $\theta_{\max} = 2\pi$ for a search of the densest possible structure. But empirically it was found that the densest possible structures for $1 \leq D \leq 1 + 1/\sin(\pi/5)$ as predicted from simulated annealing could already be recovered if θ_2 is varied only from 0 to $\theta_{\max}/2 = \pi$. We thus define our packing algorithm by this reduced range of θ_2 as adopted in the present search procedure, though one can always expand it into the whole range of 2π. Each subsequent sphere, including the third sphere, is also to be placed at a position of smallest z, determined via a 2π angular scan which starts from the angular position of the previous sphere and which goes along the direction of increasing θ. If there exist multiple positions of smallest z, the sphere is to be placed at the first of such positions encountered in the angular scan. The packing process resembles a process of sequential deposition where the spheres are "dropped" one by one onto their "lowest" positions of smallest z. The simultaneous deposition of

$u > 1$ spheres, as discussed in Section 4.1.3, does not concern this particular regime of D because we cannot place multiple spheres at the same z position. On the other hand, if a sphere is placed at a position with a z value smaller than that of any previous sphere, the resulting columnar structure of spheres would be discarded and the search for the densest possible structure would be continued with other values of θ_2. This is merely a computational shortcut to achieve a faster search for densest possible structures at the expense of some search space and also to facilitate a more convenient evaluation of (N/L). It does not affect the outcome of the search.

4.4.2 Packing Algorithm for $D \geq 2$

For $D \geq 2$, if θ_2 is equal to or greater than a critical angle given by the value of $|\Delta\theta|$ for two spheres in contact at $|\Delta z| = 0$ [Equation (4.6)], the second sphere can be placed on the flat base of the cylinder, i.e., $z_2 = 0$, after the first sphere is placed at $z = 0$ and $\theta = 0$. In order to remove any possible geometric influence from the flat base, the z position, z_2, of the second sphere also needs to be varied for a search of the densest possible structure. In this case, the search would become a two-dimensional one. A range of $z_2 \in [0, 1]$ is adopted for such a search, which turns out to be enough for recovering the densest possible structures at $1 \leq D \leq 1 + 1/\sin(\pi/5)$. If the value of D is sufficiently large such that the cylinder can accommodate a third or more spheres on its flat base after the deposition of the second sphere, additional rules based on a continuity of relative displacements of successive spheres are used to determine the positions of such spheres. These successive spheres and the first two spheres are regarded as spheres that make up a "template" for the deposition process. By definition, the creation of such a template is considered completed only if no subsequent spheres can potentially be placed onto the flat base at $z = 0$.

The additional rules for such template creation are as follows: For $n > 2$, the angular position of the nth sphere is given by $\theta_n = (n-1)\theta_2$, and this sphere is to be placed at a position of smallest z; however, if the sphere ends up on the flat base again, i.e., at $z = 0$, the continuity rule should be applied to its z position as well, namely $z_n = (n \quad 1)z_2$. In other words, the sphere would only be

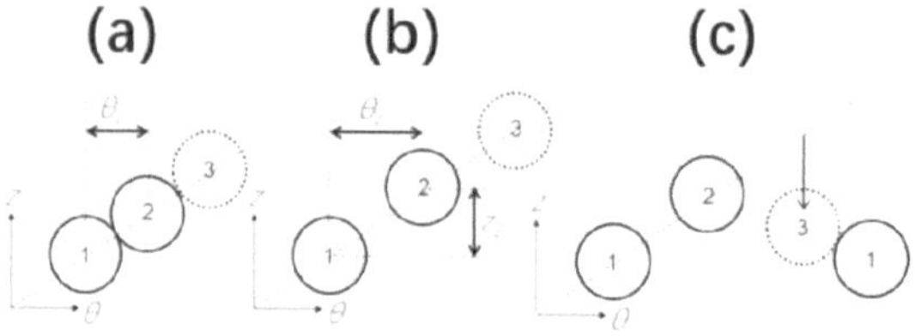

Figure 4.11 Schematic illustration of the rules for template creation at $D \geq 2$, where the continuity of relative displacements is illustrated by the dotted straight lines: (a) If θ_2 is small enough such that, at its position of smallest z, the second sphere would come into contact with the first one, θ_2 is the only parameter to be varied. (b) But if θ_2 is larger than a critical angle such that the second sphere would fall onto the flat base, the z position of the second sphere has to be assigned separately. (c) The third and subsequent spheres would follow the continuity of relative displacements along the angular dimension, but for the z dimension they would be placed at a position of smallest but non-zero z (i.e., a z position at which they are "supported" by some other spheres) wherever possible.

"elevated" to z_n if it finds itself resting on the flat base, otherwise it would just stay at its position of smallest z. These additional rules, as illustrated schematically in Figure 4.11, were inspired by the surface representations of densest possible structures at $D \geq 2$: They mimic the packings of circular disks on a plane, where in most cases densest-packed hexagonal-lattice-like regions of circular disks appear to have slid against each other to fulfill a certain periodicity requirement [Mughal *et al.* (2011, 2012); Mughal and Weaire (2014)], as illustrated in Figure 4.12.

In those regions, there exists a continuity of relative displacements of adjacent disks along each of the three crystallographic axes, i.e., the sphere centres can be joined by a straight line, and therefore it is not unreasonable to suppose that such continuity would exist for the present sphere-packing problem wherever possible. For each deposition step, the rules described above also apply to a simultaneous deposition of $u > 1$ spheres into the same z position at regular angular spacings, provided that the value of D is no smaller than the value of $D_{\min,u}$ as given by Equation (4.22). The simulation array now corresponds to an angular range of $\theta_{\max} = 2\pi/u$, and for any value of D the angular position θ_2 of the second sphere is varied from 0 to $\theta_{\max}/2$.

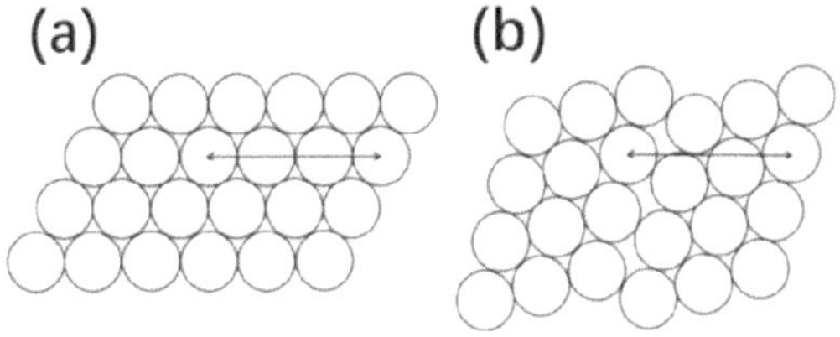

Figure 4.12 Dense packings of circular disks at a given periodicity. (a) If the periodicity is a multiple of the sphere diameter, the disks will be packed into a perfect hexagonal layer, as described by a hexagonal lattice, for a densest possible configuration. (b) If this periodicity is decreased a little bit such that it is no longer a multiple of the sphere diameter, the spheres would have to be moved around to accommodate the new periodicity, where the solution shown corresponds to a sliding of densest-packed hexagonal-lattice-like regions of circular disks.

4.5 Columnar Structures from Sequential Deposition

In this section, we present a few examples of columnar structures as obtained from the abovementioned rules for a deposition of $u = 1$ sphere at each deposition step. We first discuss columnar structures at $D = 2.35$, followed by a discussion of columnar structures at $D = 2.25$. For the latter, special attention is paid to an unconventional hybrid-helix structure that appears to be the combination of a single- and a double-helix structure.

4.5.1 Examples of Structures at $D = 2.35$

For the case of $D = 2.35$, Figure 4.13 shows how the packing fraction ϕ varies with θ_2, for a regime of θ_2 where the first two spheres are in mutual contact. For any given value of θ_2 in this regime, z_2 would just be the smallest, non-zero z position of the second sphere, and it does not have to be varied in the search for densest possible structures.

As shown in the figure, there exist three peaks for such a plot, where they correspond to three of the four structures labelled from (a) to (d). Of particular interest are structures (a) and (c) which exist as a triple helix [Figure 4.14(a)] and a quadruple

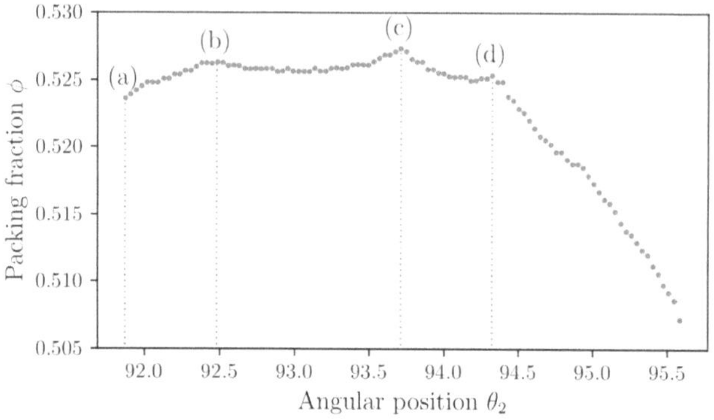

Figure 4.13 Plot of the packing fraction ϕ against the angular position θ_2 for the case of $D = 2.35$, within a regime of θ_2 where the first two spheres are in mutual contact . Four structures of interest, labelled from (a) to (d) and existing at $\theta_{2,a} \approx 91.87\,^\circ$, $\theta_{2,b} \approx 92.48\,^\circ$, $\theta_{2,c} \approx 93.71\,^\circ$, and $\theta_{2,d} \approx 94.32\,^\circ$ respectively, are identified, with structures (b) to (d) corresponding to local peaks in the plot. Of particular interest are structures (a) and (c), which exist respectively as a triple helix and a quadruple helix. Structures (b) and (d) are some intermediate structures that cannot be identified as belonging to any known type of helical structures.

helix [Figure 4.14(c)], respectively. Even though they correspond to local peaks in the plot, structures (b) and (d), as shown in Figure 4.14(b) and Figure 4.14(d), respectively, are only some intermediate structures that cannot be identified as belonging to any known type of helical structures.

The densest possible structure, however, is another triple helix (Figure 4.15) that exists in the regime of θ_2 where the value of z_2 has to be assigned and varied—its packing fraction is $\phi \approx 0.5280$, which might not be distinguishable from the value of $\phi \approx 0.5236$ for the triple helix in Figure 4.14(a) if one takes into account possible numerical uncertainties as well as the possible use of a different empirical definition of packing fraction in other contexts. It can be seen that columnar structures of equal-sized spheres as constructed from this method of sequential deposition depend sensitively on the configuration of spheres in the underlying template. Varying the angular position θ_2 of the second sphere over a range of only a few degrees could already yield a variety of structures, as demonstrated

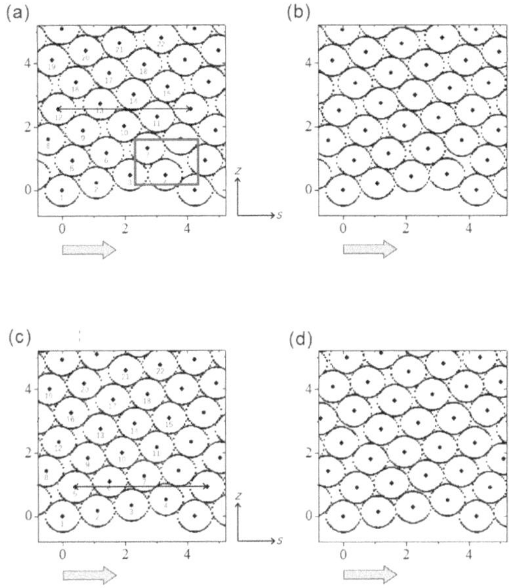

Figure 4.14 Surface representations of the four structures of interest as labelled from (a) to (d) in Figure 4.13. Structure (a) is a triple-helix structure at an angular separation of $\theta_{2,a} \approx 91.87°$ between the first two spheres; the rectangle highlights a defect-like transient configuration that is not part of the resulting periodic helical structure. Structure (c) is a quadruple-helix structure at an angular separation of $\theta_{2,c} \approx 93.71°$ between the first two spheres. For each of (a) and (c), the numbers denote the sequence of spheres in the deposition process, the arrow indicates the direction of each angular scan for positions of smallest z, and the double arrow indicates a periodicity that corresponds to the cross-sectional circumference of the inner cylindrical surface. Structures (b) and (d), which correspond to $\theta_{2,b} \approx 92.48°$ and $\theta_{2,d} \approx 94.32°$, respectively, are some intermediate structures that cannot be identified as belonging to any known type of helical structures.

in Figure 4.13 and Figure 4.14 for the case of $D = 2.35$. Such sensitive dependence makes it possible to develop a general template-based method for manipulating the dense columnar structures of identical spheres in cylindrical confinement.

As shown in Figure 4.14(a), we observe something counter-intuitive for the triple-helix structure: The fourth sphere, deposited right after the template is formed with the first three spheres, is

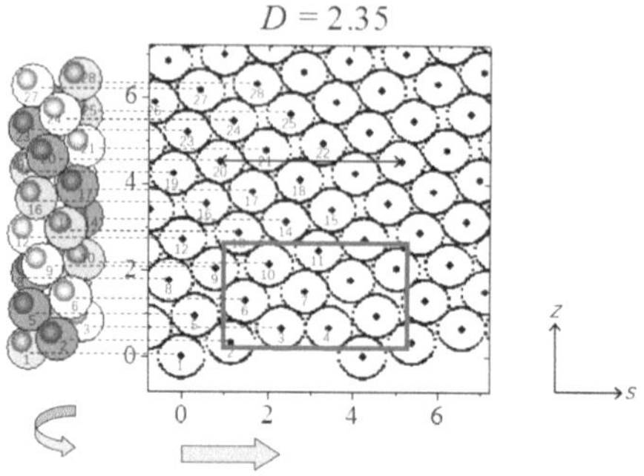

Figure 4.15 Schematic illustration of the process of sequential deposition for constructing the densest possible triple-helix structure at $D = 2.35$. The second sphere, which is not in contact with the first one, is placed at an angular position of $\theta_2 \approx 98.68°$ and a vertical position of $z_2 \approx 0.3288$ with respect to the position of the first sphere. The rectangle highlights a transient configuration that is not part of the resulting periodic helical structure.

displaced from its expected position in the periodic structure, giving rise to a defect or a transient configuration as indicated by the rectangle in the surface representation; yet, this does not obstruct the emergence of a periodic, triple-helix structure in the deposition of subsequent spheres. This defies one's expectation that all spheres must be part of the periodic structure. The same can be seen in Figure 4.15 where the transient configuration exists across a slightly wider range of z. To eliminate any possible influence from such defects on the evaluation of the packing fraction, only spheres at $z > z_c$ are used in the evaluation of (N/L). For a sufficiently long tube of $z_{max} = 15$ the value of $z_c = 2$ works well for our simulations.

4.5.2 Examples of Structures at $D = 2.25$

As for the case of $D = 2.25$, the densest possible structure (Figure 4.16) exists in a regime of θ_2 where the second sphere could be placed on the flat base. In this regime of θ_2, the position z_2 of the second sphere has to be varied from 0 to 1 in any search for the densest possible structure.

On the other hand, for the regime of θ_2 where the first two spheres are in mutual contact, there exist two peaks in the plot of the packing fraction ϕ against θ_2 (Figure 4.17). While the peak at a larger value of θ_2 for $D = 2.25$ corresponds to a conventional triple-helix structure [Figure 4.18(b)], the other peak, at a smaller value of θ_2, corresponds to an unconventional helical structure [Figure 4.18(a)] that appears to be a hybrid of the single and the double helix. This novel hybrid-helix structure, which was reported by Ho-Kei Chan [Chan (2013)], consists of three strands of spheres, with two of the strands closely packed with each other but attached in a less dense manner to a third strand.

This example demonstrates that varying the configuration of spheres in the template allows not only a manipulation of the columnar structures of spheres as constructed via the method of sequential deposition but also a possible discovery of new structures, an approach one might refer to as "template-controlled columnar crystallography".

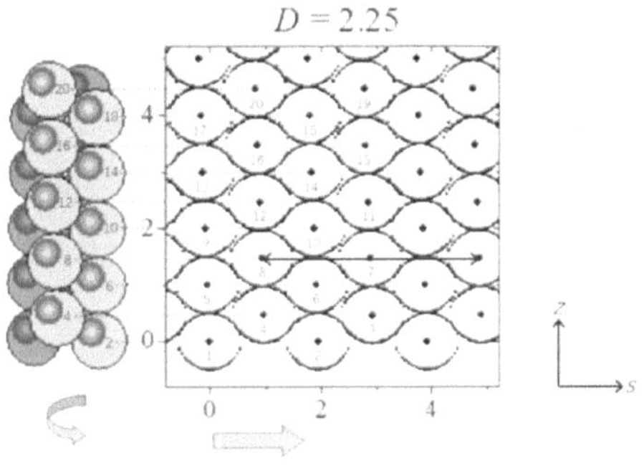

Figure 4.16 Schematic illustration of the process of sequential deposition for constructing the densest possible structure at $D = 2.25$, where its packing fraction is $\phi \approx 0.5276$. The second sphere, which is not in contact with the first one, is placed at an angular position of $\theta_2 \approx 179.7°$ and a vertical position of $z_2 \approx 0.0012$ with respect to the position of the first sphere, hence the similarity of this structure with the "achiral doublets" at $D = 2$ (Figure 4.6). Note that this structure is slightly chiral and as a result its sequence of sphere deposition is different from that of the "achiral doublets".

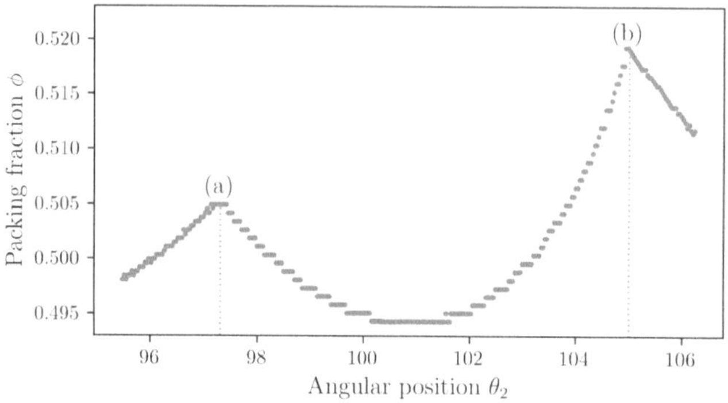

Figure 4.17 Plot of the packing fraction ϕ against the angular position θ_2 for the case of $D = 2.25$, within a regime of θ_2 where the first two spheres are in mutual contact. Two helical structures of interest, labelled as (a) and (b) and existing as a less dense hybrid helix (a hybrid of the single and the double helix) at $\theta_{2,\text{a}} \approx 97.31°$ and as a denser triple helix at $\theta_{2,\text{b}} \approx 105°$, respectively, are identified.

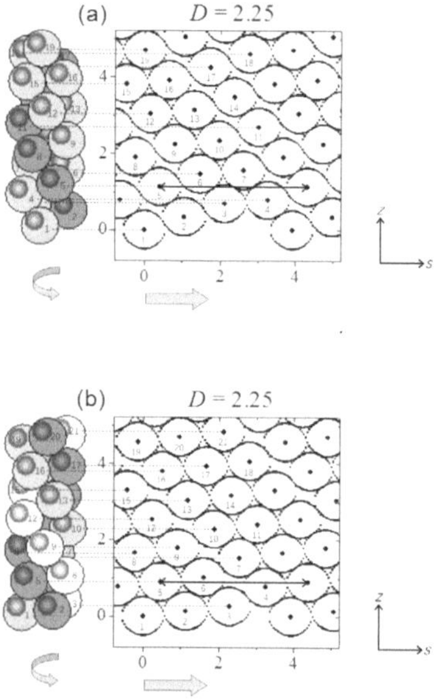

Figure 4.18 Schematic illustration of the process of sequential deposition for constructing (a) a hybrid helix and a (b) triple helix, at $D = 2.25$, which are referred to as structures (a) and (b), respectively. In the unconventional hybrid-helix structure, two strands of spheres are closely packed with each other but they are attached in a less dense manner to a third strand of spheres, thus mimicking a hybrid of the single and the double helix. In both cases, the first two spheres are in mutual contact.

4.6 Conclusions

In this chapter, we describe in detail a method of sequential deposition for constructing the densest possible structures of equal-sized hard spheres at $D \in [1, 1 + 1/\sin(\pi/5)]$. For any densest structure in this regime of D, all spheres are in contact with the wall of the confining cylinder. This method of sequential deposition is essentially a surface-packing algorithm, which can be extended beyond $D = 1 + 1/\sin(\pi/5)$ to generate dense packings of spheres on the curved surface of a cylinder, regardless of what the densest columnar structures of spheres are in this extended regime of D. This constitutes an interesting problem of surface packing in its own rights, where it would be worthwhile to find out how the system undergoes a transition to the expected two-dimensional hexagonal-lattice structure at $D \to \infty$ for increasing D.

By dropping spheres one by one into their lowest possible positions, the packing fraction of a vertically aligned columnar structure is maximised locally during the addition of each sphere. However, this does not guarantee the resulting columnar structure to be the densest possible. It is because the overall structure can be "fine-tuned" by adjusting the positions of the first few spheres at the bottom, i.e., the configuration of the underlying template. For example, by perturbing the second sphere away from its lowest possible position at an angular displacement of π from the first sphere, there would be additional space to accommodate the third sphere at a lower vertical position [Chan *et al.* (2019)]. The density of the triplet of spheres is optimised at the expense of the density of the first two spheres. The same applies iteratively to all consecutive triplets of spheres (i.e., spheres $\{2,3,4\}$, $\{3,4,5\}$, $\{4,5,6\}$...), resulting in an increase of the overall packing fraction of the columnar structure.

On the other hand, such sensitive dependence of any resulting structure on the configuration of the underlying template turns out to be a blessing in disguise. While for any given value of D an exhaustive search across a variety of template configurations is required to obtain the densest possible columnar structure, some novel crystal structures show up unexpectedly during such a search. A well-studied example is the unconventional hybrid-helix structure at $D = 2.25$ [Chan (2013)]. This opens up new avenues for the development of "template-controlled columnar crystallography", for which we expect some more novel crystal structures to be discovered upon an exhaustive variation of template configurations.

Chapter 5

Soft-Sphere Packings in Cylinders

Relevant journal publications by J. Winkelmann:

- A. Mughal, J. Winkelmann, D. Weaire and S. Hutzler, Phys. Rev. E 98, 043303 (2018)
- J. Winkelmann, B. Haffner, D. Weaire, A. Mughal and S. Hutzler, Phys. Rev. E 97 059902 (2018).

In this chapter, we present a computed phase diagram of columnar structures of soft spheres under pressure. The diagram is characterised by the first appearance and then disappearance of line-slip structures, as the pressure is increased. A comparable experimental observation was made on a column of bubbles under forced drainage, clearly exhibiting the expected line slip. This constituted the first experimental realisation of a line-slip structure in a soft-sphere system. Since macroscopic systems of this kind are not confined to ideal equilibrium states, some of the structural transitions that are to be expected as experimental conditions are varied have also been explored, where such transitions are in general hysteretic. Such transitions are best represented in a stability diagram, of which we present a representative case. These results are then extended to include all

Columnar Structures of Spheres: Fundamentals and Applications
Jens Winkelmann and Ho-Kei Chan

ISBN 978-981-4669-48-1 (Hardcover), 978-0-429-09211-4 (eBook)
www.jennystanford.com

stable structural transitions between columnar structures without inner spheres (i.e., packings in which all of the spheres are in contact with the cylindrical boundary). In the hard-sphere limit, such transitions are accompanied by a loss (or gain) of contact, with the system transforming from a uniform structure to a line-slip structure (or vice versa). These results can be summarised by a directed network that indicates permissible uniform-to-uniform transformations from an initial uniform structure, over a line slip, to a final uniform structure.

5.1 Introduction to Soft-Sphere Packings in Cylindrical Confinement

Columnar structures arise when spheres are packed densely inside (or on the surface of) a cylinder [Pickett *et al.* (2000); Koga and Tanaka (2006); Duran-Olivencia and Gordillo (2009); Chan (2011); Mughal *et al.* (2011); Srebnik and Douglas (2011); Mughal *et al.* (2012); Chan (2013); Mughal (2013); Wood *et al.* (2013); Mughal and Weaire (2014); Yamchi and Bowles (2015); Fu *et al.* (2016); Fu (2017); Fu *et al.* (2017); Mughal *et al.* (2018); Chan *et al.* (2019)]. In Subsection 3.2.1, we discuss structures of hard-sphere packings as obtained in this way. At $1 \leq D \leq 2.71486$, all spheres of any densest hard-sphere packing are in contact with the cylindrical wall, and all the corresponding columnar structures have been identified and tabulated [Mughal *et al.* (2012)] (see Table A.1 of Appendix A).

The densest possible uniform structures of equal-sized hard spheres in cylindrical confinement exist at specific values of D, where each of such structures is characterised by a phyllotactic notation (l, m, n) (An introduction to phyllotactic notations is provided in Section 2.3). At all other values of D, the corresponding densest possible uniform structures present themselves in the form of line slips. Each of such line-slip structures corresponds to a shear of two adjacent spirals of a uniform structure along one of the crystallographic directions, with a corresponding loss of contacts. Both types of structures are discussed in detail in Subsection 3.3.

Since line-slip structures were first discovered in hard-sphere systems, we have become accustomed to thinking of line-slip structures as a property of *hard*-sphere packings and therefore judging them to be of limited relevance to real physical systems. That point of view was reconsidered by Winkelmann *et al.* [Winkelmann *et al.* (2018)]. In this chapter, we demonstrate the experimental existence of line-slip structures in wet foams with high liquid fractions. Discounting the trivial case of packing ball-bearings in tubes [Mughal *et al.* (2012)], this constitutes the first ever experimental observation of line-slip structures. This experimental work was stimulated in part by the observation of line slips (albeit rather indistinctly) in simulations where microscopic particles interact via Lennard–Jones-type potentials [Wood *et al.* (2013); Douglass *et al.* (2017)].

Previous computational investigations of hard-sphere packings have been extended to packings of soft repelling spheres, where such soft spheres are allowed to overlap. As discussed in Section 5.2, such overlapping is penalised by the increase of an interaction energy that scales with the square of the overlap. The resulting columnar structures depend not only on the diameter ratio D but also on the applied uniaxial pressure P. A phase diagram that displays the stable structures of lowest enthalpy at different values of D and P has also been obtained from such simulations.

In addition, simulations results from an exploration of metastability and hysteresis are discussed in this chapter. *Given a locally stable structure and an (experimental) protocol for a continuous variation of D, what types of structural transitions are to be expected?* This question has been addressed by investigating a particular example of a reversible structural transition in detail and presenting the results in the form of a stability diagram. With results from further investigations, the full picture of all possible structural transitions of uniform structures is displayed in the form of a directed network.

5.2 Simulations: Minimisation of Enthalpy H

To simulate packings of soft spheres in cylindrical confinement, the *soft*-sphere model by Durian [Durian (1995)] was employed (see Subsection 3.2.2). For sphere i and sphere j with their centres located at $\vec{r}_i$ and $\vec{r}_j$ respectively, the overlap $\delta_{ij} = d_{\text{sphere}} - |\vec{r}_i - \vec{r}_j|$

corresponds to an interaction energy

$$E_{ij}^{S} = \begin{cases} 0 & \delta_{ij} < 0 \\ \frac{1}{2}k\delta_{ij}^{2} & \delta_{ij} \geq 0 \end{cases} \tag{5.1}$$

For any sphere i at a radial distance r_i from the central axis, the harmonic energy

$$E_i^{B} = \frac{1}{2}k\left(\frac{D_{\text{cylinder}} - d_{\text{sphere}}}{2} - r_i\right)^2 \tag{5.2}$$

accounts for any overlap between the sphere and the cylindrical boundary. For both types of interactions, the softness of a sphere is determined by the spring constant k.

Here, we describe simulations using a unit cell of length L, diameter D_{cylinder}, and volume $V = \pi(D_{\text{cylinder}}/2)^2L$, with a total of N spheres in the unit cell (Figure 5.1). Screw-periodic columnar structures of soft spheres have been obtained through the use of *twisted* periodic boundary conditions at both ends of the unit cell. These conditions were implemented by placing image spheres both above and below the unit cell. Each image sphere corresponds to

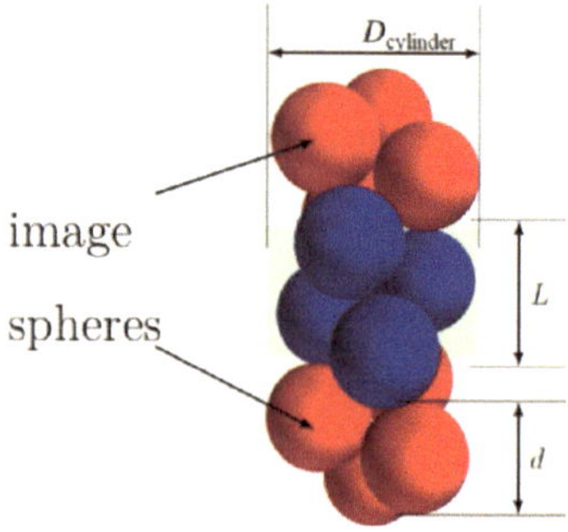

Figure 5.1 Simulations of N soft spheres (blue) of diameter d_{sphere} in cylindrical confinement. The unit cell of the structure has a diameter D_{cylinder} and a length L. Twisted periodic boundary conditions, with a vertical separation L and an angular separation α between each image sphere and its real counterpart, are implemented at both ends of the unit cell.

a vertical displacement from a real sphere by L or $-L$ and to an angular displacement from the same real sphere by α or $-\alpha$ along the horizontal plane. In this method, only interactions among real spheres, and interactions between real and image spheres, are taken into account for the total energy of the simulation cell. Interactions among image spheres are disregarded in the definition of the total energy.

For a system with a total of N soft spheres in its unit cell, stable structures have been obtained by a minimisation of the enthalpy $H = E + PV$, where E is the internal energy due to overlaps between spheres, P is the pressure, and V is the volume.

The internal energy $E = E^{S'} + E^{S} + E^{B}$ is made up of energy $E^{S'}$ for interactions between real spheres and image spheres, energy E^{S} for interactions among real spheres, and energy E^{B} for interactions between real spheres and the cylindrical wall. The enthalpy H is thus given by

$$H(\{\vec{r}_i\}, L) = \underbrace{\frac{1}{2}\sum_{\substack{i,j=0\\ i\neq j}}^{N} E_{ij}^{S'} + \frac{1}{2}\sum_{\substack{i,j=0\\ i\neq j}}^{N} E_{ij}^{S} + \sum_{i=0}^{N} E_i^{B}}_{E} + \underbrace{PV}_{\text{pressure term}} . \tag{5.3}$$

In the minimisation of H at a constant pressure P and a variable volume, the $(3N + 2)$ free parameters are the positional coordinates $\{\vec{r}_i\}$ of the N spheres, the twist angle α, as well as the length L of the unit cell.

Alternatively, this problem can be viewed as a problem of constrained minimisation where the internal energy E is minimised subject to the constraint of a constant pressure. The quantity to be minimised is $(E - \lambda P)$, where λ is the Lagrange multiplier. This turns out to be the volume V of the system.

Throughout this chapter, we adopt some terminology from thermodynamics, such as enthalpy, pressure and phase diagram, even though the system under study is athermal. This is justified by the observation of thermodynamic-like effects such as discontinuous and continuous transitions as well as hysteresis in such an athermal system.

Two types of minimisation routines have been employed to minimise the enthalpy $H(\{\vec{r}_i\}, \alpha, L)$. For densest possible packings as indicated by the phase diagram, we search for the global minimum of the enthalpy. For the stability analysis of any given structure and for any investigation of corresponding structural transitions, we only search for the nearest local minimum.

The BFGS method (named after Broyden, Fletcher, Goldfarb, Shanno [Byrd *et al.* (1995)]) is used to search for the nearest *local* minimum of enthalpy. It is a quasi-Newton algorithm that searches iteratively for the nearest minimum from an initial guess, until the absolute value of the gradient falls below a tolerance threshold.

The Basin-hopping algorithm [Wales and Doye (1997)] performs a general search for the *global* minimum. It is a stochastic algorithm that searches iteratively for the global minimum of the enthalpy. Each iteration consists of three steps: (1) Random perturbation of the coordinates; (2) Local minimisation; (3) Acceptance or rejection of the new coordinates. For any computational step to be accepted, the Metropolis criterion is employed. The above-mentioned BFGS method is used to perform the step of local minimisation.

These two algorithms, together with a variety of other minimisation algorithms, are introduced in Section 3.5 and explained in detail in Appendix B.

In the simulations, the dimensionless enthalpy $h \equiv H/(kd_{\text{sphere}}{}^2)$ and the dimensionless pressure $p \equiv P/(k/d_{\text{sphere}})$ are considered, where k is the spring constant and d_{sphere} is the sphere diameter. It follows from the definition of p that the hard-sphere limit at $k \to \infty$ corresponds to the limit of $p \to 0$ [Mughal *et al.* (2012)] (see Subsection 3.2.1). The validity of such soft-sphere simulations can be verified by checking for any agreement between soft-sphere results at $p \to 0$ and existing results for hard-sphere packings.

Alternatively, simulations can also be carried out via a minimisation of the system's internal energy at a fixed volume, i.e., at a fixed value of L. Then, however, there are three independent control parameters in each simulation: the diameter ratio D, the fixed length L, and the softness k of the spheres. Such simulations correspond to a different experimental scenario with a more complicated three-dimensional phase diagram. The simulations discussed in this chapter only involved the approach of enthalpy minimisation.

5.3 Simulation and Observation of Line-Slip Structures in Soft-Sphere Packings

With the simulation method described above, columnar structures of soft-sphere packings can be generated at different values of the diameter ratio $D \equiv D_{\text{cylinder}}/d_{\text{sphere}}$ and the dimensionless pressure p. By systematically generating structures at given values of D and p, a phase diagram of all stable, lowest-enthalpy structures without internal spheres has been obtained. As discussed in Section 3.3, these structures are classified into two types, namely *uniform* structures and *line-slip* structures.

5.3.1 Phase Diagram of All Uniform and Line-Slip Structures without Internal Spheres

Figure 5.2 presents a phase diagram for all *uniform* and *line-slip* arrangements at $1.5 \leq D \lesssim 2.7$ and $p \leq 0.02$. The lower limit of $D = 1.5$ was chosen arbitrarily. This is justified since only the trivial bamboo structure at $D = 1$ (where the spheres are stacked on top of each other) exists below this limit. The upper limit of $D \sim 2.7$ corresponds to a transition to structures with the presence of internal spheres [Mughal *et al.* (2012)].

At $p > 0.02$, all line-slip structures vanish. At large pressures, the contact area between any two overlapping spheres becomes comparable to the dimensions of a sphere, and the model of soft-sphere packings may be regarded as unrealistic. As p increases, a system of foam bubbles approaches the "dry limit" at which the liquid fraction tends to zero. This is discussed in detail in Subsection 2.4.2.

For any given values of D and p, the optimal structure as indicated in the phase diagram was obtained by minimising the enthalpy of an N-sphere unit cell. The structure with the lowest enthalpy was chosen. At low pressures, the process of minimisation was carried out via the Basin-hopping algorithm. These low-pressure results were used as initial guesses (i.e., starting points for enthalpy minimisation) in the BFGS method for mapping out the higher-pressure regions of the phase diagram. Starting with some initial structure at $N = 3$, $N = 4$ or $N = 5$, the pressure p and the diameter ratio D were

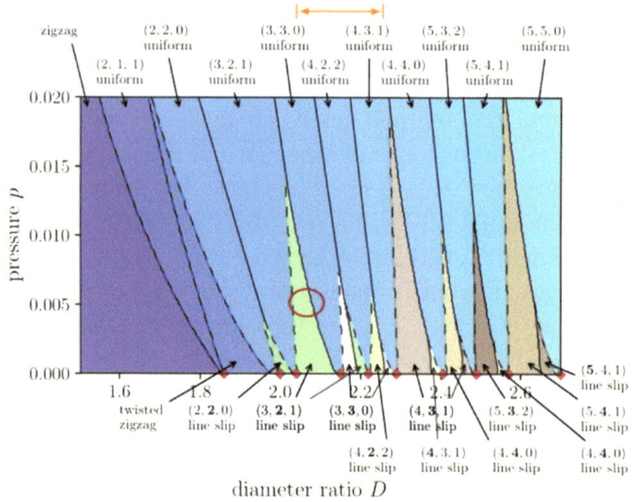

Figure 5.2 Phase diagram for soft-sphere packings in cylindrical confinement at diameter ratios of $1.5 \leq D \lesssim 2.7$ and dimensionless pressures $p \leq 0.02$. The resolutions along D and p are respectively $\Delta D = 0.0025$ and $\Delta p = 0.0004$. Apart from the zigzag structure, the $(2, 1, 1)$ uniform structure and the single-helix (twisted-zigzag) structure, there exist ten uniform (blue shaded) structures and fourteen related line-slip structures (green and brown). Small regions that indicate the $(\mathbf{2}, 1, 1)$ and the $(\mathbf{3}, 2, 1)$ line-slip structures (as observed for hard-sphere packings) are not visible in this phase diagram, as a result of the corresponding resolution limits. Discontinuous transitions are indicated by solid black lines, while continuous transitions are indicated by black dashed lines. The diamond symbols at $p = 0$ indicate the uniform structures at the hard-sphere limit [Mughal *et al.* (2012)]. The orange arrow right above the phase diagram indicates the range of D where a more detailed investigation for different possible types of structural transitions was conducted (see Subsection 5.3.2). A rough estimate of an experimentally observed line-slip structure (see Subsection 5.3.3) is indicated by the ellipse on this phase diagram.

increased independently. To check the validity of these results, the two parameters were also varied in a different manner: Starting with some initial structure at a high value of D, the value of D was reduced in discrete steps while keeping the pressure p constant. For each pair of parameter values, the corresponding lowest-enthalpy structure is obtained and then indicated in the phase diagram.

For $D \geq 2$, a total of twenty-four distinct structures, including ten uniform structures and fourteen related line-slip structures, have been obtained. For $D < 2$, the bamboo structure, the zigzag structure, the $(2, 1, 1)$ uniform structure, as well as the $(2, \mathbf{1}, 1)$ line-slip structure have been observed.

5.3.2 Structural Transitions in the Phase Diagram

The transitions between different structures in the phase diagram can be classified as follows: In any *continuous* structural transition (as indicated by a dashed line in Figure 5.2), a structure transforms smoothly into another structure by gaining or losing a contact. This is shown in the supplemental video S0 of a paper by Mughal *et al.* [Mughal *et al.* (2018)], where the video presents an overview of all the structures at a relatively low pressure together with the corresponding rolled-out contact networks as well as the location of each structure in the phase diagram. A continuous structural transition occurs between a uniform structure and a line slip, e.g., between the $(3, 2, 1)$ uniform $\rightarrow$ $(3, \mathbf{2}, 1)$ line slip transition or between the zigzag structure $\rightarrow$ $(2, 1, 1)$ uniform structure.

In any *discontinuous* structural transition (as indicated by a solid line in Figure 5.2), a structure with a particular phyllotactic notation (l, m, n) is transformed into another structure with a completely different phyllotactic notation (i, j, k). Such a transition may also be described as abrupt (see the supplementary video S0 by Mughal *et al.* [Mughal *et al.* (2018)]). A typical example is the discontinuous transition between the line-slip structures of $(3, \mathbf{2}, 1)$ and $(3, \mathbf{3}, 0)$. For increasing pressure, the regions of both line-slip structures narrow, with that of the $(3, \mathbf{2}, 1)$ line slip disappearing first at a triple point. This is followed by the disappearance of the region of the $(3, \mathbf{3}, 0)$ line slip at a second triple point. At pressures above this second triple point, the uniform structures of $(3, 3, 0)$ and $(4, 2, 2)$ are separated by a discontinuous structural transition.

It appears from the phase diagram of Figure 5.2 that the region between the $(3, 2, 1)$ and $(3, 3, 0)$ uniform structures is exceptional. Only a single line slip region for the $(3, \mathbf{2}, 1)$ is visible. According to Table A.1 [Mughal *et al.* (2012)] of Appendix A, however, there should be the presence of a second line slip. The $(\mathbf{3}, 2, 1)$ line-slip structure is present within the narrow range of $2.1413 \leq D \leq 2.1545$ for hard-sphere packings. This line-slip structure is not visible in this computed phase diagram due to the corresponding resolution limit of the diagram.

The distinction between a continuous and a discontinuous structural transition can be illustrated directly via the dimensionless enthalpy h. In the following, we focus on a particular range of D, which is indicated by an orange arrow in the phase diagram of Figure 5.2.

Figure 5.3(a) shows a plot of the dimensionless enthalpy h against D at a constant pressure of $p = 0.01$. In a continuous structural transition, such as that between the $(3, 2, 1)$ uniform structure (green) and the $(3, \mathbf{2}, 1)$ line-slip structure (light-green region), the variation of h with D is smooth such that the transition point is not apparent. In a discontinuous structural transition, such as that between the $(3, \mathbf{2}, 1)$ line-slip structure and the $(3, 3, 0)$ uniform structure (yellow region), there is an abrupt change in the slope $\partial h/\partial D$. Figure 5.3(b) shows a corresponding plot of the compression $C \equiv (V_0 - V(p, D))/V_0$ as a function of the diameter ratio D. For any

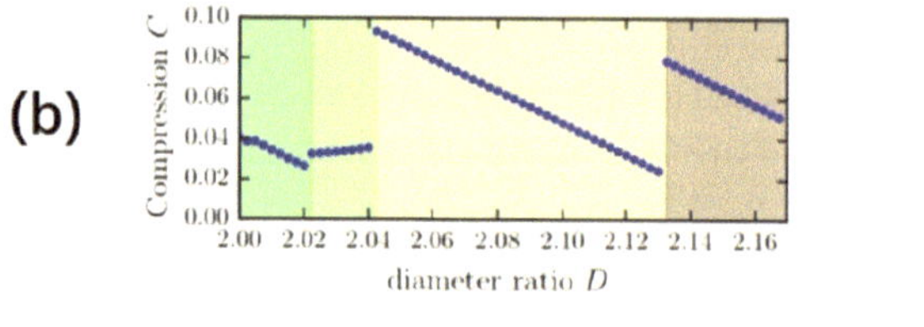

Figure 5.3 (a) Plot of the dimensionless enthalpy h as the diameter ratio D is varied at a constant pressure of $p = 0.01$. The plot corresponds to a horizontal cut through the phase diagram for the chosen range of D as indicated by the orange arrow. (b) Plot of the compression C of soft-sphere packings as a function of D. Only the structural transition between the $(3, 2, 1)$ uniform structure and the $(3, \mathbf{2}, 1)$ line-slip structure is continuous, as characterised by a smooth variation of h.

given values of p and D, $V(p, D)$ is the unit-cell volume of the soft-sphere packing, and V_0 is a unit-cell volume of the corresponding columnar structure of hard spheres. For any uniform structure of hard spheres, the value of V_0 is unique [Mughal (2013)]. For any line-slip structure of hard spheres, however, the length and volume of the unit cell generally vary with D. We take V_0 to be the volume of the smallest possible unit cell for any particular line-slip structure [Mughal (2013)].

It is also worthwhile to consider a vertical cut through the phase diagram, i.e., variation of dimensionless enthalpy h as a function of p at a constant diameter ratio D. Figure 5.4(a) shows a plot of h as a function of p at $D = 2.1$. The corresponding plots of $\partial h/\partial p$ and C as functions of p are shown in Figure 5.4(b)

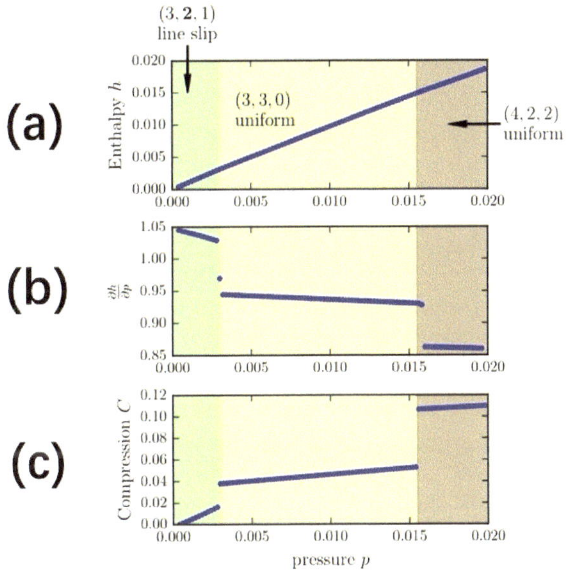

Figure 5.4 (a) Plot of the dimensionless enthalpy h as the dimensionless pressure p is varied at a constant diameter-ratio of $D = 2.1$. (b) Plot of the numerically computed derivative $\partial h/\partial p$ as a function of p. (c) Plot of the compression C of soft-sphere packings as a function of p. The two structural transitions shown in the figure are discontinuous, with each transition characterised by a discontinuity in the derivative $\partial h/\partial p$. The plots in (a) and (b) correspond to a vertical cut through the phase diagram of Figure 5.2.

and Figure 5.4(c), respectively. For increasing p, there exist two discontinuous structural transitions, one between the $(3, \mathbf{2}, 1)$ line-slip structure and the $(3, 3, 0)$ uniform structure, and another between the uniform structures of $(3, 3, 0)$ and $(4, 2, 2)$. While the abrupt change in the slope of h at each transition is not obvious in Figure 5.4(a), it is confirmed by the plot of $\partial h/\partial p$ in Figure 5.4(b). Note the similarities between the derivative in enthalpy (b) and the compression (c): The enthalpy in Figure 5.4(b) increases whereas the compression in Figure 5.4(c) decreases with discontinuities at the transition points. This allows us to classify any transition as continuous or discontinuous in terms of the thermodynamic concepts of first- and second-order transitions.

5.3.3 Experimental Observation of Line-Slip Structures

In the previous section, we show that line-slip structures exist not only for packings of hard spheres but also for those of soft spheres. Figure 5.2 demonstrates the existence of line-slip structures up to a certain pressure (or overlap). But so far line-slip structures of soft particles have only been obtained from computer simulations.

These simulations have led to a prediction of line-slip structures for experimental systems of soft (deformable) spheres. Figure 5.5 illustrates the existence of line-slip structures for soap bubbles inside a glass tube. By comparing the capillary pressure of a bubble with its hydrostatic pressure, it was shown that bubbles packed inside a tube can be approximated as soft spheres, thus confirming the possible experimental existence of line-slip structures for soft particles.

All the experiments described here were performed by Dr. Benjamin Haffner, a former post-doc in the Foams and Complex Systems research group of Trinity College Dublin. The experimentally created foam structures were identified and analysed by Jens Winkelmann.

Columnar structures of bubbles have been obtained under *forced drainage*, i.e., through a steady input of liquid from the top of a system [Weaire *et al.* (1993, 1997); Koehler *et al.* (2000)]. This method is explained in detail in Subsection 2.4.2. In the past, columnar structures of foams at much larger values of D have been studied extensively via this method. It was found that convective

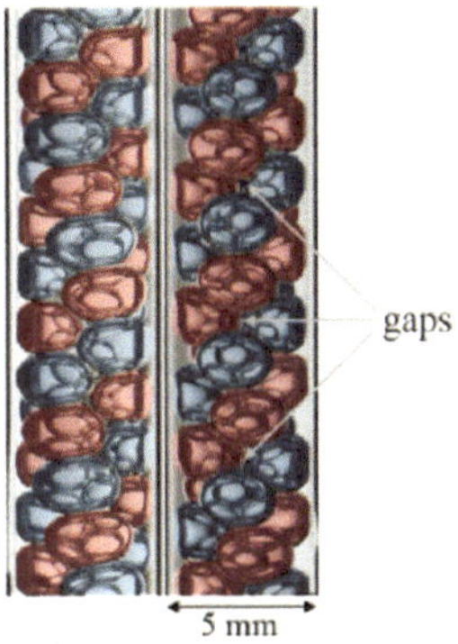

Figure 5.5 Columnar structures of about forty bubbles under forced drainage. **Left:** A $(3, 2, 1)$ uniform structure appears at a relatively low flow rate. **Right:** A $(3, \mathbf{2}, 1)$ line-slip structure, which is similar to the line-slip structure shown in Figure 3.4(b), appears at a relatively high flow rate. The post-processed colouring highlights the two interlocked spirals of each structure. In the phase diagram of Figure 5.2, the estimated location of this line-slip structure is marked by a red ellipse.

instabilities occur at flow rates that lead to a relatively high liquid fraction. Research was therefore restricted to relatively dry foams [Hutzler *et al.* (1998)]. In this chapter, we consider columnar structures of wet foams where near-spherical bubbles are produced at a sufficiently high flow rate of liquid without the occurrence of convective instability in the chosen experimental range of D. Ordered columnar structures that resemble the structures obtained from simulations have been produced. The liquid fraction of a foam column is controlled by a variation of the flow rate.

The experimental set-up as introduced in Subsection 2.4.2 was employed. Monodisperse bubbles were produced by blowing air through a needle into a solution of the commercial surfactant Fairy Liquid [Weaire *et al.* (1993); Hutzler *et al.* (1997)]. The solution contains 50 % of glycerol by weight, where the presence of glycerol would increase the viscosity of the solution, smoothen the transition between structures, and stabilise the arrangements of bubbles. The gas-flow rate ($q_0 \sim 1$ mL/min) was adjusted to produce monodisperse bubbles of a particular size. The diameter $d_{\text{sphere}} \sim (2.50 + 0.04)$ mm of a bubble was determined as follows:

A monolayer of bubbles was produced by confining a small amount of foam within a Hele-Shaw-like set-up, i.e., between two horizontal, parallel plates separated by a negligibly small gap. The cross-sectional area A of each bubble was determined via image analysis. The diameter of each bubble was then given by $d_{\mathrm{sphere}} = \sqrt{2A/\pi}$.

After being injected into a vertically aligned cylindrical glass tube of inner diameter 5 mm and length 1.5 m, the bubbles self-organised into an ordered columnar structure. The resulting foam column was put under forced drainage by feeding from the top a surfactant solution, with the liquid flow rate Q being up to $\sim$ 10 mL/min. The capillary pressure was estimated to be $p_{\mathrm{c}} = \gamma/r \sim 10\mathrm{Pa}$, where $\gamma = 0.03$ N/m and $r = d_{\mathrm{sphere}}/2$ are the surface tension and the radius of a bubble, respectively. For bubbles at a vertical position x, the hydrostatic pressure is given by $p_{\mathrm{h}} = \rho g x$, where ρ is the density of water and g is the gravitational acceleration. At $x \sim 1$ m, the hydrostatic pressure is $p_{\mathrm{h}} \sim 10^4\mathrm{Pa}$, three orders of magnitude larger than the capillary pressure, and is therefore a dominant contribution. The bubbles in this pressure regime can easily be deformed, and the foam can safely be regarded as a packing of relatively *soft* objects.

Figure 5.5 displays an experimentally observed $(3, \mathbf{2}, 1)$ line-slip structure. The loss of contacts in this experimental structure resemble that of the simulated $(3, \mathbf{2}, 1)$ line-slip structure from the red-ellipse region of Figure 5.2. The experimental values of the pressure (determined from the local packing fraction) and diameter ratio are consistent with estimates from the phase diagram. For the relatively short experimental column of bubbles in Figure 5.5, the pressure difference across the column is negligibly small. If such an investigation is extended to the full length of the tube, the resulting column of bubbles will display a spatial variation of structures as a result of a pressure gradient along the vertical direction. We discuss possible experiments of this type in Section 5.5.

While the experimentally observed $(3, \mathbf{2}, 1)$ line-slip structure is a stable structure in the red-ellipse region of Figure 5.2, it can also exist as a metastable structure outside this region. It is therefore necessary to extend these investigations of soft-sphere packings to the context of metastability and hysteresis.

5.4 Hysteresis and Metastability in Structural Transitions

For microscopic systems, the phase diagram in Figure 5.2 applies only to stable structures in equilibrium. For some systems, such as those currently under investigation [Yin and Xia (2003); Tsytovich *et al.* (2007); Zerrouki *et al.* (2008); Lohr *et al.* (2010); Fu *et al.* (2017); Wu *et al.* (2017)], many of the observed structures are metastable, i.e., they are not lowest-enthalpy structures. Pittet *et al.* also discovered that the structural transitions are not necessarily reversible: An original structure of soap bubbles is not necessarily recovered, when the variation in the control parameter is reverted [Pittet *et al.* (1995); Boltenhagen *et al.* (1998)]. That is, columnar structures can be metastable and their structural transitions are in general hysteretic.

A wider context of metastable structures and hysteretic structural transitions has been explored through further simulations. This enables a comprehensive theoretical description of all possible structural transitions as well as the region of (meta-)stable structures. These are both vital information for the experimental realisation of columnar structures.

Where previously the computation of a phase diagram entailed a search for the *global* minimum, involving the Basin-hopping algorithm that allowed radical changes of structures to be explored, here we pursue a more limited objective: *Given a stable or metastable initial structure, how does the structure change when the diameter ratio D (and/or the pressure p) are varied, if it is to remain in the nearest local minimum of enthalpy (to the initial structure)?* Since we are only interested in a local minimum of enthalpy, the BFGS method was employed. To ensure that the nature of hysteresis investigated in this section is independent of the minimisation method, a subset of results was corroborated with the gradient descent routine (for both methods, see also Section 3.5.1).

Highly symmetric structures, such as the $(4, 2, 2)$ uniform structure, might get stuck at a saddle point where the gradient is zero but the enthalpy is not minimal. The system can be moved out of the saddle point by *small* random perturbations of the spheres' positions, where such perturbations typically do not exceed 10% of the spheres' diameter.

Simulations were carried out for a unit cell with $N = 12$ spheres. This chosen value of N is suitable for structures with one, two, three or four spheres in their primitive unit cells (see Table A.1 of Appendix A), and is sufficient for the considered range of D. Structures with five spheres in their unit cells at relatively high values of D are outside the scope of this chapter.

In the following section, we present computed trajectories of local minima of enthalpy in the (p, D) parameter space and we record the boundaries at which a structure is transformed to a different structure. With a sufficient number of trajectories, a *stability diagram* is created, i.e., a map that indicates locations of structural transitions in the parameter space.

Here, we describe in greater details the structural transitions between the uniform structures of $(3, 2, 1)$ and $(4, 2, 2)$ as well as transitions associated with the $(3, \mathbf{2}, 1)$ line-slip structure. We begin our discussion by showing how the enthalpy h varies with the diameter ratio D at a constant pressure p. We show that, at low pressures, it is possible to start with any of the above-mentioned structures and continuously transform one structure into another through a change in D. In such a transformation, a uniform structure is transformed into a line-slip arrangement through a loss of contacts or a line-slip structure is transformed into a uniform structure through a formation of contacts. This is generally accompanied by a continuous variation in h. At high pressures, however, some of these transitions are no longer reversible and they exhibit hysteretic behaviour. Such discontinuous transformations are accompanied by a discontinuity in h. From these results, a stability diagram is obtained from which a supplementary schematic stability diagram is extracted.

We also present a directed network for all the uniform structures indicated in the phase diagram of Figure 5.2. The network displays *all* permissible structural transitions between uniform structures without internal spheres and summarises all possible types of structural transitions of uniform structures. These transitions are not necessarily reversible, as in the case of the $(3, 2, 1) \leftrightarrow (4, 2, 2)$ transition.

5.4.1 Enthalpy Curves at Constant Pressures for a Reversible Transition

An example of a computed enthalpy curve is shown in Figure 5.6. It demonstrates how the enthalpy h varies for increasing D, at a constant pressure of $p = 0.007$. At such a low pressure, an increase in D allows the $(3, 2, 1)$ uniform structure to transition continuously into the $(3, \mathbf{2}, 1)$ line-slip structure through a loss of contacts, and then continuously into the $(4, 2, 2)$ uniform structure through a formation of new contacts, with both transitions being reversible. The dashed vertical lines in the figure indicate the values of D where the structural transitions occur. The smooth change in the enthalpy over the course of these transitions implies that the process is continuous and reversible. The smooth variation of structures, together with the corresponding rolled-out contact networks and the corresponding locations in the stability diagram of Figure 5.9(b), are shown in the supplemental video S1 by Mughal *et al.* [Mughal *et al.* (2018)]. Note that information of these transitions cannot be obtained from the phase diagram (Figure 5.2), which is based on global minima of enthalpy. The phase diagram predicts the otherwise

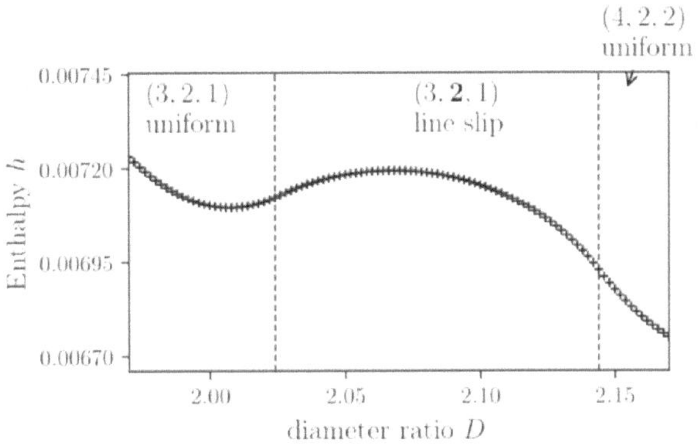

Figure 5.6 Dimensionless enthalpy h as a function of D at a relatively low pressure of $p = 0.007$. At such a low pressure, an increase in D allows the $(3, 2, 1)$ uniform structure to transition continuously into the $(3, \mathbf{2}, 1)$ line-slip structure through a loss of contacts, and then continuously into the $(4, 2, 2)$ uniform structure through a formation of new contacts, with both transitions being reversible. The dashed vertical lines in the figure indicate the values of D at which the structural transitions occur.

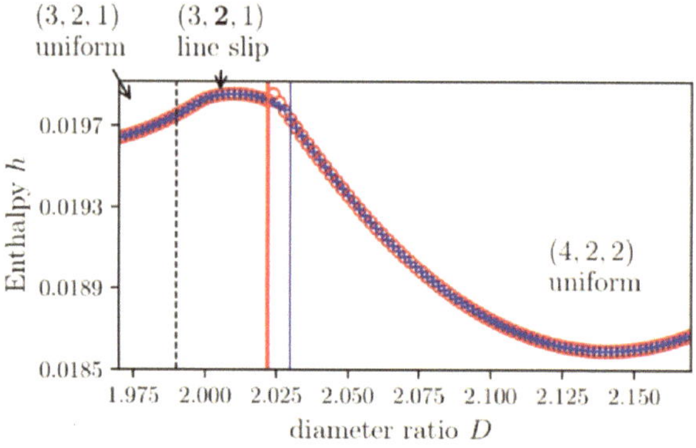

Figure 5.7 Dimensionless enthalpy h as a function of D at a higher pressure of $p = 0.020$, for increasing D (blue crosses) and decreasing D (red circles). A variation in D results in a continuous and reversible transition at the vertical dashed line between the $(3, 2, 1)$ uniform structure and the $(3, \mathbf{2}, 1)$ line-slip structure. However, the transition between the $(3, \mathbf{2}, 1)$ line-slip structure and the $(4, 2, 2)$ uniform structure is discontinuous, where the transition for decreasing D (vertical red solid line) occurs at a smaller value of D than the transition for increasing D (vertical blue solid line).

occurrence of the $(3, 3, 0)$ uniform structure, which is different from the scenario in Figure 5.6.

At a higher pressure of $p = 0.02$, the nature of structural transitions is different, as illustrated in Figure 5.7. For increasing D (blue crosses), a transition from the $(3, 2, 1)$ uniform structure to the $(3, \mathbf{2}, 1)$ line-slip structure occurs at the vertical dashed line, with a smooth variation in the enthalpy h as in the case of the low-pressure example described above. However, a further increase in D results in a discontinuous transition into the $(4, 2, 2)$ uniform structure at the vertical blue solid line, with an observable drop in the slope of h (see the supplemental video S2 by Mughal *et al.* [Mughal *et al.* (2018)]). Upon a decrease in D (red circles), the transition from the $(4, 2, 2)$ uniform structure to the $(3, \mathbf{2}, 1)$ line slip is also discontinuous. However, it occurs at a smaller value of D as indicated by the vertical red solid line. The results suggest the existence of a critical pressure, above which a hysteresis of structural transitions occurs.

At a relatively high pressure of $p = 0.026$, the $(3, \mathbf{2}, 1)$ line-slip structure disappears completely for decreasing D, as illustrated

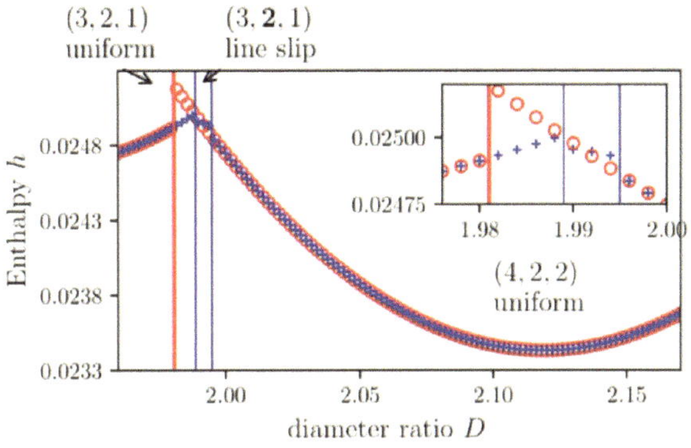

Figure 5.8 Dimensionless enthalpy h as a function of the diameter ratio D at a relatively high pressure of $p = 0.026$, for increasing D (blue crosses) and decreasing D (red circles). For increasing D, a discontinuous transition between the $(3, 2, 1)$ uniform structure and the $(3, \mathbf{2}, 1)$ line-slip structure occurs at the vertical blue solid line on the left, and a discontinuous transition between the $(3, \mathbf{2}, 1)$ line-slip structure and the $(4, 2, 2)$ uniform structure occurs at the vertical blue solid line on the right. For decreasing D, the intervening $(3, \mathbf{2}, 1)$ line-slip structure disappears, such that the $(4, 2, 2)$ uniform structure transforms to the $(3, 2, 1)$ uniform structure at the vertical red solid line in a direct discontinuous transition. The inset is a zoom-in view of these discontinuous transitions.

in Figure 5.8. The $(4, 2, 2)$ uniform structure transforms directly to the $(3, 2, 1)$ uniform structure by a discontinuous transition (vertical red solid line) without the presence of an intervening line-slip structure. For increasing D, the transition between the $(3, 2, 1)$ uniform structure and the $(3, \mathbf{2}, 1)$ line-slip structure, and that between the $(3, \mathbf{2}, 1)$ line-slip structure and the $(4, 2, 2)$ uniform structure, are both discontinuous. A further increase of the pressure to $p \gtrsim 0.028$ eliminates the $(3, \mathbf{2}, 1)$ line-slip structure for increasing D as well, such that there exist only discontinuous transitions between the uniform structures of $(3, 2, 1)$ and $(4, 2, 2)$ (see the supplemental videos S3 and S4 by Mughal *et al.* [Mughal *et al.* (2018)]).

5.4.2 Stability Diagram for a Reversible Transition

A stability diagram represents the boundaries at which transitions among a specific set of structures take place. Such a diagram only displays transitions that start from one of the structures considered. There may be the existence of some other metastable structures that are not indicated in the diagram. As discussed in a previous paragraph, a stability diagram must not be confused with a phase diagram.

Figure 5.9(a) shows a stability diagram for transitions among the uniform structures of $(3, 2, 1)$ and $(4, 2, 2)$ as well as the $(3, \mathbf{2}, 1)$ line-slip structure. It was obtained from a computation of the corresponding enthalpy curves (Subsection 5.4.1). Figure 5.9(b) is

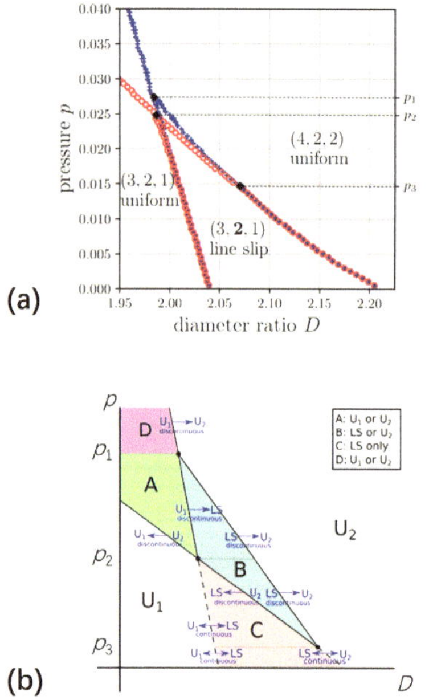

Figure 5.9 (a) Stability diagram for transitions among the uniform structures of $(3, 2, 1)$, $(4, 2, 2)$ as well as the $(3, \mathbf{2}, 1)$ line-slip structure. The transition points for increasing D and for decreasing D are indicated by the blue crosses and red circles, respectively. (b) Schematic stability diagram representing the qualitative features of the stability diagram (a). The two uniform structures are labelled as U_1 and U_2, respectively, and the intermediate line-slip structure is labelled as LS. Each dashed line represents a boundary for continuous reversible transitions, and each solid line represents a boundary for discontinuous irreversible transitions. The arrows at each line corresponds to the direction, to which the transition occurs.

a schematic diagram that serves as a supplementary interpretation of this stability diagram. It represents the qualitative features of the stability diagram but does not preserve the quantitative features. The symbol U_1 corresponds to the $(3, 2, 1)$ uniform structure, the symbol U_2 to the $(4, 2, 2)$ uniform structure, and the symbol LS represents the $(3, \mathbf{2}, 1)$ line-slip structure. Each dashed line represents a boundary for continuous reversible transitions, and each solid line represents a boundary for discontinuous irreversible transitions. The blue arrows indicate the allowed directions for the corresponding structural transitions.

Figure 5.9 identifies four distinct pressure regimes that correspond to the examples described above. These are $p < p_3$ (Figure 5.6), $p_3 < p < p_2$ (Figure 5.7), $p_2 < p < p_1$ (Figure 5.8) and $p_1 < p$. For the regime of $p_1 < p$, the $(3, \mathbf{2}, 1)$ line-slip structure has completely vanished. In this figure, each blue cross represents a transition for increasing *D*, and each red circle represents a transition for decreasing *D*. The supplemental videos S1 to S4 by Mughal *et al.* [Mughal *et al.* (2018)] illustrate the variation of structures for increasing or decreasing *D* at a constant pressure.

As an example, we discuss in detail the case of $p = 0.020$ $(p_3 < p < p_2)$, which corresponds to the enthalpy curves in Figure 5.7. The other cases can be analysed in a similar manner by comparing Figure 5.9(a) with Figure 5.9(b).

Starting from the $(3, 2, 1)$ uniform structure at $p = 0.020$, an increase in *D* results in a continuous transition to the $(3, \mathbf{2}, 1)$ line-slip structure, as indicated by the blue crosses in Figure 5.9(a). In Figure 5.9(b), this corresponds to the dashed line for the continuous transition $U_1 \leftrightarrow$ LS.

A further increase in *D* results in a discontinuous transition (blue crosses) to the $(4, 2, 2)$ uniform structure. In Figure 5.9(b), this is indicated by a solid line and by the notation LS $\rightarrow U_2$. The blue arrow indicates that this discontinuous transition only occurs for increasing *D*.

For decreasing *D*, the $(4, 2, 2)$ uniform structure undergoes a discontinuous transition (red dots) into the $(3, \mathbf{2}, 1)$ line-slip structure. In Figure 5.9(b), this is indicated by a solid line and by the notation LS $\leftarrow U_2$. A further decrease of *D* results in a continuous transition into the original uniform $(3, 2, 1)$ structure, at the boundary marked by red dots in Figure 5.9(a) and a dashed line in Figure 5.9(b). We have thus returned to the starting point of this particular excursion through the stability diagram.

5.4.3 Directed Network of Structural Transitions

The stability diagram in Figure 5.9 provides detailed information about possible reversible transitions, i.e., transition of U_1 into U_2 for increasing D ($U_1 \rightarrow U_2$), and transition of U_2 back into U_1 for decreasing D ($U_1 \leftarrow U_2$). However, this is only part of the picture, as there exist other types of structural transitions. The directed network in Figure 5.10 summarises all possible structural transitions for the uniform structures in the phase diagram of Figure 5.2.

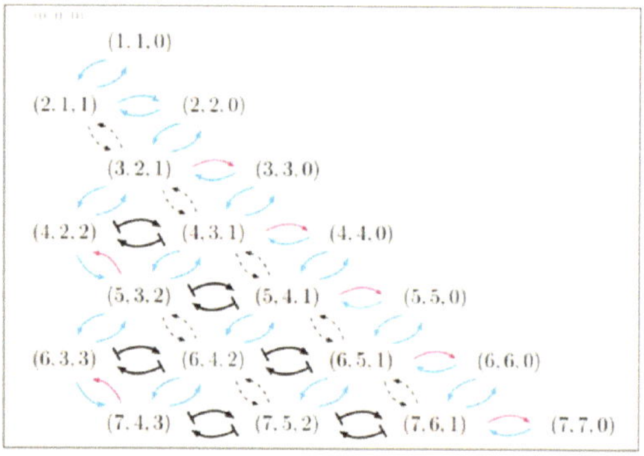

Figure 5.10 Directed network that summarises all possible structural transitions for the uniform structures in the phase diagram of Figure 5.2. The point $(0, 0, 0)$ represents a conceptual origin that corresponds to $D = 0$. The dashed lines are contours of constant D. If a structural transition points in a direction perpendicular to the contours, there must be a corresponding change in D. Favourable transitions are indicated by blue solid arrows, and unfavourable transitions by red solid arrows. The black solid arrows and black dashed arrows denote some other structural transitions that are not discussed in this monograph. We have identified a counterclockwise cyclic chain of favourable structural transitions: $(3, 2, 1) \leftrightarrow (4, 2, 2) \rightarrow (5, 3, 2) \leftrightarrow (4, 3, 1) \leftrightarrow (3, 3, 0) \rightarrow (3, 2, 1)$. The two red solid arrows that point from $(3, 2, 1)$ and $(5, 3, 2)$ prohibit this cyclic chain of structural transitions from going clockwise.

To construct such a network, we addressed the following question by considering the hard-sphere limit of soft-sphere simulations: *Starting from a given uniform structure, which structural transitions are favourable upon a variation of the diameter ratio D?*

For any given uniform structure, there might exist multiple intermediate line-slip structures, each of which would lead to a unique uniform structure. In the simulations, only a subset of possible structural transitions have been observed, to which we will refer as *favourable* transitions.

Together with existing information on unfavourable transitions of hard-sphere packings (see Figure A.1 and Figure A.2 in Appendix A), a directed network that displays all possible structural transitions of uniform structures has been worked out.

We consider a conceptual origin $(0, 0, 0)$: The closer a structure is to this origin, the smaller is the corresponding value of D. The dashed lines represent contours of constant D about the origin of $D = 0$. If a structural transition points in a direction perpendicular to the contours, there must be a corresponding change in D.

In Figure 5.10, favourable transitions are indicated by blue solid arrows, and each blue-arrow pair corresponds to a reversible transition [e.g., the $(3, 2, 1) \leftrightarrow (4, 2, 2)$ transition]. Red solid arrows indicate transitions that are not favourable. For decreasing D, a transition of the $(3, 3, 0)$ uniform structure to the $(3, 2, 1)$ uniform structure is favourable, but the reverse transition for increasing D is not favourable.

For the sake of clarity, all intermediate line-slip structures are left out. Each arrow represents the transition from a particular uniform structure, over some line-slip structure, to another uniform structure with a different phyllotactic notation. Similar investigations can also be carried out for softer spheres or at a higher pressure, where all line-slip structures disappear if the spheres are sufficiently soft or the pressure sufficiently high. But the directed network in Figure 5.10 for uniform structures remains the same.

Following the favourable transitions in the graph, cyclic chains that can only be followed counterclockwise are identified. An example that is located closest to the conceptual origin of the directed network starts with the $(3, 2, 1)$ structure: $(3, 2, 1) \Leftrightarrow (4, 2, 2) \Rightarrow (5, 3, 2) \Leftrightarrow (4, 3, 1) \Leftrightarrow (3, 3, 0) \Rightarrow (3, 2, 1)$. The two unidirectional

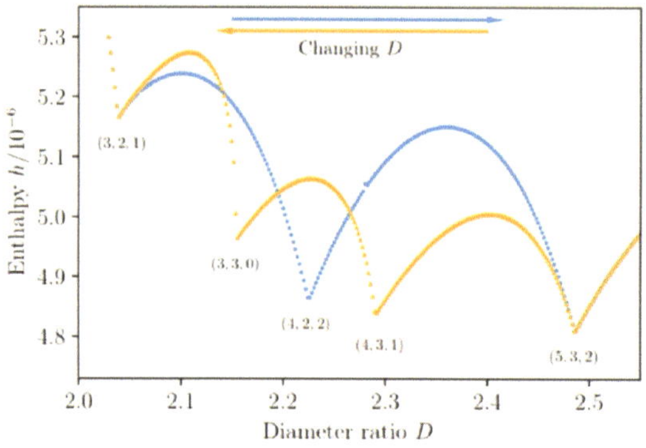

Figure 5.11 Plot of the enthalpy h for the counterclockwise cycle that is closest to the conceptual origin in Figure 5.10 as a function of the diameter ratio D at constant pressure $p = 5 \times 10^{-6}$. The cycle starts from the local minimum that corresponds to the $(3, 2, 1)$ uniform structure. The blue dots indicate the enthalpy values of the favoured intermediate line-slip structures for increasing D. This trajectory leads the initial $(3, 2, 1)$ uniform structure over the $(4, 2, 2)$ into the $(5, 3, 2)$ uniform structure. Decreasing D from there (yellow triangles), a different intermediate line slip is preferred, leading to a different uniform structure at the minima of the yellow enthalpy curve. However, by decreasing D further, the $(3, 2, 1)$ uniform structure is reached again, such that the cycle is completed.

arrows at $(3, 2, 1)$ and $(5, 3, 2)$ prevent this cycle from going clockwise.

Figure 5.11 displays two enthalpy curves h (one for increasing D, the other for decreasing D) for the above-mentioned cyclic chain of favourable structural transitions at a relatively low pressure, i.e., close to the hard-sphere limit. Each uniform structure corresponds to a local minimum of enthalpy. All other parts of the curves correspond to enthalpy values of intermediate line-slip structures. For increasing D (blue dots), the $(3, 2, 1)$ uniform structure undergoes a transition to the $(4, 2, 2)$ uniform structure via an intermediate line-slip structure. This is followed by a transition to the $(5, 3, 2)$ uniform structure via another line-slip structure. Decreasing D at this point leads us back to the (3, 2, 1) uniform structure through a different enthalpy curve (yellow triangles), via the path of $(5, 3, 2) \to (4, 3, 1) \to (3, 3, 0) \to (3, 2, 1)$. This counter-clockwise cycle implies that a transformation from the $(3, 2, 1)$ uniform structure to the $(3, 3, 0)$ uniform structure cannot be achieved by *simply* increasing D, even though they exist as adjacent structures in the phase diagram of Figure 5.2.

From the directed network, we can also identify all reversible structural transitions, where such transitions are denoted by double blue arrows. Examples include the transitions of $(3,2,1) \Leftrightarrow (4,2,2)$, $(4,3,1) \Leftrightarrow (5,3,2)$, $(5,3,2) \Leftrightarrow (6,3,3)$, ... The behaviour of such transitions is qualitatively similar to that of the examples described in Section 5.4. Therefore, they can all be described by the schematic stability diagram in Figure 5.9(b).

A similar directed network by Fu *et al.* for hard-sphere packings is shown in Figure 3.6 [Fu *et al.* (2017)]. While the directed network in Figure 5.10 is consistent with that in Figure 3.6, it provides much more informative predictions about possible transitions for increasing as well as decreasing D and allows us to identify cyclic chains of favourable transitions. Another related study is that by Pittet *et al.* [Pittet *et al.* (1996)], who investigated structural transitions of *dry* foams in cylindrical confinement. Starting from the simple $(1,1,0)$ bamboo structure of a linear chain of bubbles, Pittet *et al.* identified two distinct sequences of transitions for increasing D. Their findings were presented in a similar grid of structural transitions [Pittet *et al.* (1996)]. However, they have discovered a different set of favourable transitions since the soft-sphere model does not describe a dry foam system.

5.5 Conclusions

We have shown that line-slip structures exist not only for hard-sphere packings but also for soft-sphere packings. Line-slip structures are therefore relevant to a variety of soft matter systems, such as foams, emulsions, and colloids. The results presented in this chapter, in particular the phase diagram in Figure 5.2, may be compared with earlier studies [Wood *et al.* (2013); Wu *et al.* (2017)] that adopted the Lennard-Jones potential for inter-particle interactions. In these studies, both uniform structures and line-slip structures were observed. On the other hand, for very soft particles, such as those interacting via the Yukawa potential, line-slip structures have not been predicted to exist [Oğuz *et al.* (2009)], nor have they been observed in experiments [Wu *et al.* (2017)].

This absence of line-slip structures is consistent with theoretical predictions from the phase diagram in Figure 5.2.

For soft-sphere packings, all the structures indicated in Figure 5.2 also exist for hard-sphere packings. No new structure has been discovered in the studies of soft-sphere packings. It is possible that there exist new structures at pressures beyond the scope of this research. However, at such high pressures, the soft-sphere model employed in this research becomes unrealistic, and some other models such as the Z-cone model [Hutzler *et al.* (2014)] or the Morse–Witten model [Dunne *et al.* (2019)] (Chapter 8) might be better candidates.

We have also presented studies of soft-sphere packings in the context of metastability and hysteresis, where the corresponding results are presented in the form of a stability diagram (Figure 5.9). For structures indicated in the phase diagram of Figure 5.2, a directed network (Figure 5.10) that summarises all possible structural transitions of uniform structures has been constructed. All reversible structural transitions, displayed in this network, can be described by the stability diagram in Figure 5.9(a) and the schematic stability diagram in Figure 5.9(b).

It is perhaps a surprise that all structures in the stability diagram of Figure 5.9 are only structures that can be found in the phase diagram of Figure 5.2, albeit over different ranges of p and D. The lack of newly discovered structures is expected at or close to the hard-sphere limit, but might not have been anticipated for higher pressures. The employed minimisation procedure does not exclude any possible discovery of novel structures, for instance structures with defects and/or kinks [Figure 2.10(c) in Subsection 2.4.3]. It is not clear why none of such structures have been obtained in the simulations.

For future work, it would be worthwhile to extend this analysis of hysteresis and metastability to higher values of D, for which Fu *et al.* [Fu *et al.* (2016)] have computed a list of equilibrium structures that are of a different character (Section 3.4). This is likely to be quite demanding, and should perhaps be guided by some preliminary experimental investigations.

On the other hand, these results should also provide insights for new experiments in which it is more convenient to vary p for fixed D, rather than the converse. A possible set-up for such experiments is the bubble column described in Subsection 5.3.3, where there already exists a gravity-induced vertical variation of p. Such a system may be used to identify structures and transitions that can be compared with the ones presented in this chapter. Diagrams such as Figure 5.9(b) may then serve as a guide to the practical fabrication of structures of soft spheres in tubes, for which the phase diagram of Figure 5.2 by itself may be misleading due to the hysteretic nature of corresponding structural transitions.

Chapter 6

Rotational Columnar Structures of Soft Spheres

Relevant journal publications by J. Winkelmann:

- J. Winkelmann, A. Mughal, D. B. Williams, D. Weaire and S. Hutzler, Phys. Rev. E 99, 020602(R) (2019)

By introducing a new assembly method, Lee et al. used rapid rotations around a central axis to drive spheres of a lower density than the surrounding fluid towards this axis [T. Lee, K. Gizynski, and B. Grzybowski, Adv. Mater. ***29****, 1704274 (2017)]. This resulted in different structures as the number of spheres is varied. In this chapter, we present comprehensive analytic energy calculations for such self-assembled structures, based on a generic soft-sphere model, from which we obtain a phase diagram. It displays interesting features, including peritectoid points. These analytic calculations are complemented by preliminary numerical simulations for finite-sized systems of soft spheres. A similar analytic approach could be used to study packings of spheres inside cylinders of fixed dimensions, but with a variation in the number of spheres.*

Columnar Structures of Spheres: Fundamentals and Applications
Jens Winkelmann and Ho-Kei Chan

ISBN 978-981-4669-48-1 (Hardcover), 978-0-429-09211-4 (eBook)
www.jennystanford.com

6.1 Introduction to Rotational Columnar Structures

In 2017, the subject of columnar structures took a new turn thanks to the development of a novel experimental method by Lee *et al.* [Lee *et al.* (2017)]. In their experiments, the rapid spinning of a liquid-filled column, which contains spheres of a lower density than the surrounding fluid, drives the spheres towards the central axis. Confined by a centripetal force, the spheres self-assemble into a columnar structure around this axis.

Lee *et al.* performed lathe experiments with the use of polymeric beads, which are impenetrable and essentially behave like hard spheres. They have observed a variety of columnar structures that depend on the number of spheres and the speed of rotation. Their experimental findings were corroborated by molecular dynamics (MD) simulations.

A similar approach can also be adopted for columnar structures obtained as such with *soft* spheres. This chapter will display analytic methods to calculate the energy of such structures for soft spheres. From these analytic results, we obtained a comprehensive phase diagram, with the two parameters of the diagram being the linear number density (number of spheres per unit length) and the rotational frequency. The phase diagram is packed with great features, such as *peritectoid points*, that we will discuss in detail in Subsection 6.3.2.

We have also compared such analytic results to soft-sphere simulations of finite-sized systems. This comparison revealed some surprising modifications to the analytic phase diagram of infinitely long systems.

Lee *et al.*'s simulations are computationally intensive because they describe the full motion of spheres inside a rotational flow, including inertia. On the other hand, the simulations of finite-sized systems, where the principle of energy minimisation is employed to produce equilibrated columnar structures, are computationally inexpensive. They, however, do not describe accurately the dynamics of the system's evolution towards equilibrium.

6.2 Lee *et al.*'s Lathe Experiments

In this section, we summarise the methodology and related results of Lee *et al.*'s innovative experimental assembly method of creating columnar structures via rapid rotations. They suspended monodisperse spheres inside a denser rotating fluid. Rapid rotations of the system drive the spheres towards the axis of rotation, where the spheres self-assemble to a variety of ordered structures. Due to such rotations, the centripetal force confines the spheres inside a harmonic potential, where the magnitude of this force can be tuned by a variation of the rotational speed.

Figure 6.1 shows a schematic illustration of Lee *et al.*'s experimental set-up. Identical polymeric beads of diameter $d_{\text{sphere}} = 1.588$ mm and density ρ were immersed in an aqueous solution of agarose and caesium bromide. For different experiments, the density of beads was chosen over a range of $\rho = 0.9$ g/cm to $\rho = 1.13$ g/cm. By adding the compound caesium bromide, the density of the fluid was raised above the density of the beads, such that the beads became buoyant within the liquid. Inside a commercial

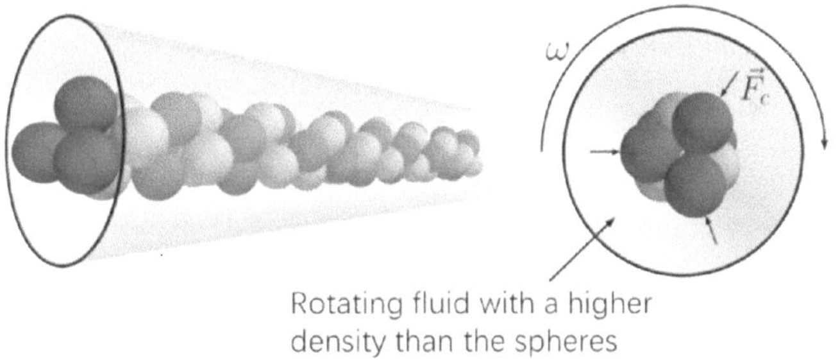

Figure 6.1 Experimental set-up for assembling columnar structures of spheres inside a rotating liquid-filled tube. The liquid has a higher density than the spheres, such that the spheres are buoyant within the liquid. When the system is rotated with a rotational speed ω, a centripetal force $\vec{F}_c$ pushes the spheres towards the axis of rotation. The spheres self-assemble around this axis into a variety of columnar structures. Figure reproduced after Lee *et al.*'s experimental description [Lee *et al.* (2017)].

lathe that rotated at a speed of 10000 rpm, the beads were pushed by a centripetal force towards the central axis, and the liquid was pushed towards the wall of the tube. Due to a relatively large Young's modulus of $E = 1325$ MPa [DesignerData (2019)], the beads are impenetrable and interact like hard spheres.

Different structures interconverted upon a variation of the speed of rotation. At a fixed tube length, the resulting structure was determined by the number of polymeric beads inside the cylindrical tube, i.e., by the number density of spheres in the system. With the agarose, the assembled structures were solidified by gelation and turned into permanent structures after the assembly process. The assembly process was captured using a high-speed camera.

To gain better insights into the assembly process, Lee *et al.* complemented their experiments with molecular dynamics (MD) simulations of (essentially) hard spheres. These simulations modelled the motion of spheres within a rotational flow, where inertial forces, drag forces in a viscous fluid, buoyancy, and sphere–sphere interactions were taken into account in the simulations. Hydrodynamic interactions were not considered.

Depending on the rotational speed ω and the number of spheres, two different types of columnar structures have been observed in experiments and simulations. A columnar structure can either be (a) a homogeneous structure with the same configuration of spheres filling out the entire tube, or (b) a binary mixture of two different configurations separated by an interface. Figure 6.2 shows two examples for each type of structures, where in Figure 6.2(b) the interface of the binary structure is highlighted.

The existence of binary mixtures seems to be surprising on first sight, because the Mermin–Wagner theorem [Mermin and Wagner (1966)] states that spontaneous symmetries cannot be broken for structures in one or two dimensions at finite temperatures. In Chapter 3, we show that the columnar structures considered in this research are essentially two-dimensional. The degrees of freedom for each cylinder-touching sphere are only its height z and its azimuthal angle θ. Following the Mermin-Wagner theorem, one may conclude that the existence of binary mixtures is prohibited in Lee *et al.*'s experiments. However, for the structures considered in this chapter, each sphere has three degrees of freedom, as it is allowed to move radially as well. This accounts for the existence of binary mixtures in

(a)

(b)

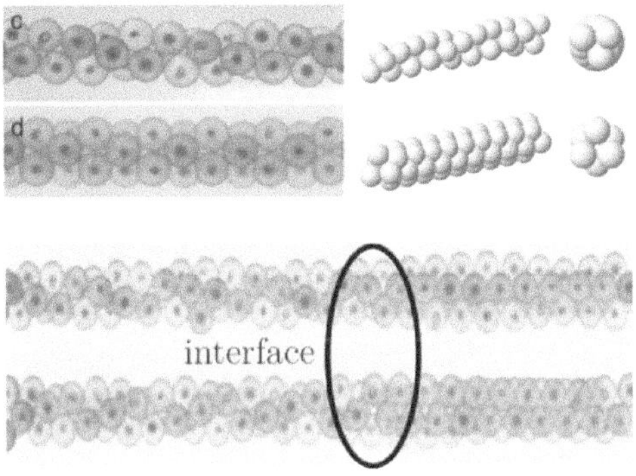

Figure 6.2 Examples of observed columnar structures in Lee *et al.*'s lathe experiments that involved rapid rotations [Lee *et al.* (2017)]. Depending on the number of spheres and the rotational speed ω, the polymeric beads self-assembled around the rotating axis either (a) as a homogeneous structures or (b) as a binary mixture of two types of homogeneous structures. A homogeneous structure consists of a single phase with a single configuration filling out the entire tube. A binary mixture is made up of two different configurations separated by an interface as indicated by the black ellipse. Both figures reprinted, with permission, from Ref. [Lee *et al.* (2017)].)

Lee *et al.*'s experiments, where not all spheres are located at the same radial distance from the central axis.

As shown in Figure 6.3, a phase diagram with the rotational speed and the number density being the corresponding parameters has been obtained by Lee *et al.* [Lee *et al.* (2017)], both experimentally and computationally. The horizontal axis corresponds to the normalised particle number n/n_0, where n is the number of polymeric beads in the tube and n_0 is the number of beads required to form a linear chain inside the tube. The value of $n/n_0 = 1$ corresponds to the presence of a linear chain.

Both the experimental and the simulated phase diagram indicate a familiar sequence of columnar structures of hard spheres. The legends on the right-hand sides of Figure 6.3 display the same sequence of homogeneous structures as that for the uniform structures in Figure 3.5. However, no intervening line-slip structure has been discovered (see Subsection 3.3.2 for an explanation of line-slip structures). Instead, binary mixtures with the co-presence of two

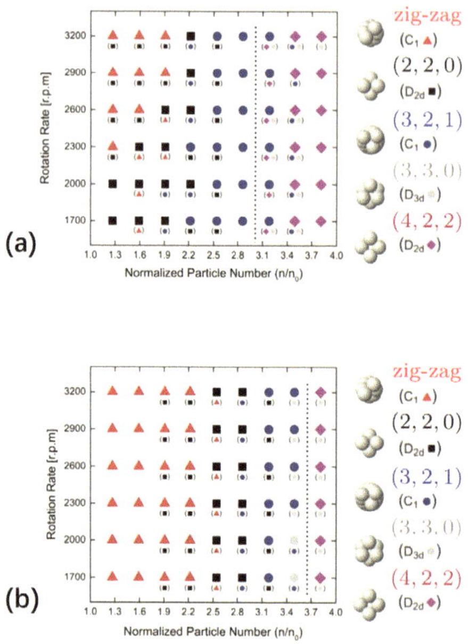

Figure 6.3 Lee *et al.*'s phase diagrams as obtained (a) from experiments and (b) from simulations, with the rotational speed and the normalized particle number n/n_0 being the corresponding parameters of each diagram. n is the number of spheres in the system, and n_0 is the number of spheres needed to form a linear chain inside the tube. Each homogeneous structure is denoted by a single symbol. For a binary mixture, the less dominant structure is indicated by a symbol in parentheses. Both figures reprinted, with permission, from Ref. [Lee *et al.* (2017)].

columnar structures for respectively different values of D have been observed.

Lee *et al.* also managed to reproduce all of their experimentally observed structures through simulations. However, a comparison between the experimental phase diagram in Figure 6.3(a) with the computational phase diagram in Figure 6.3(b) reveals a general shift towards lower number densities in the case of experiments. While this is not discussed by Lee *et al.* [Lee *et al.* (2017)], it is reasonable to suppose that vibrations induced by the spinning of the lathe are a possible origin for such a deviation between experimental and computational results. As discussed in Subsection 6.3.2, a similar shift was encountered for the analytic calculations described in this chapter.

Lee *et al.* have also discovered that the chirality of a helical structure (see Section 3.3 for a discussion of chirality in such systems) can be controlled by adjusting the orientation of the axis of rotation. While structures in a horizontally aligned tube show no preference for chirality, the beads in an inclined tube first gather at the top end and then self-assemble into a helix of a preferred type of handedness. This effect was also reproduced in their simulations.

Lee *et al.* have also experimented on columnar structures with bidisperse polymeric beads [Lee *et al.* (2017)]. The larger polymeric beads self-assembled in a chain along the axis of rotation, while the smaller beads self-assembled around the larger beads. Lee *et al.* have also presented preliminary results for beads assembled at the interface of two immiscible fluids. Such structures resemble those of spheres assembled on the surface of a cylinder.

6.3 Columnar Structures from Rapid Rotations: A Theoretical Analysis

In this section, we use the soft-sphere model to analyse Lee *et al.*'s experimental method of assembling columnar structures of spheres. We first focus on the rotational energies of hard-sphere packings (Subsection 6.3.1). Using the soft-sphere model, we then derive analytically the total energy for soft-sphere packings (Subsection 6.3.2). A phase diagram is then obtained for columnar structures without internal spheres.

6.3.1 Energy of Hard-Sphere Packings

As described in Section 3.4, Mughal *et al.* [Mughal *et al.* (2011, 2012); Mughal (2013); Mughal and Weaire (2014)] have obtained computationally a variety of densest columnar structures of hard spheres in cylindrical confinement. Despite the lack of any rigorous mathematical proof, these results are in little doubt since they have been corroborated by others [Fu *et al.* (2016); Fu (2017); Fu *et al.* (2017)]. As shown in Figure 3.5 of Section 3.4, the packing fraction

ϕ of such densest structures was computed as a function of the cylinder-to-sphere diameter ratio D. Up to $D \sim 2.7$, the spheres of such structures are all in contact with the wall of the confining cylinder, such that they are all at the same radial distance R from the central axis. This is the case for all the homogeneous structures considered here.

As discussed in Chapters 3, columnar structures of hard spheres can be classified into two types [Mughal *et al.* (2012)], namely *homogeneous* uniform structures and line-slip structures. An example for each of these types of structures is shown in the inset of Figure 6.4. For any homogeneous uniform structure, each sphere is in contact with six neighbouring spheres. The entire structure can be described in terms of a *phyllotactic* notation, i.e., a triplet of positive integers $(l = m + n, m, n)$ with $m \geq n$. (A detailed explanation of phyllotaxis is provided in Subsection 2.3.)

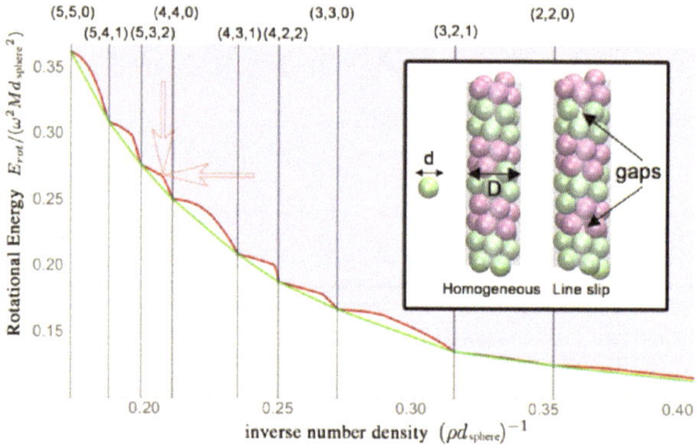

Figure 6.4 Minimal rotational energy E_{rot} as a function of the dimensionless inverse number density $(\rho d_{\text{sphere}})^{-1}$ for *hard-sphere* packings. The red solid curve corresponds to energies of line-slip structures. The vertical black solid lines indicate the locations of homogeneous uniform structures (Chapter 5) in the figure, where each of such structures is described by a particular phyllotactic notation (l, m, n). For the binary mixture between any pair of adjacent homogeneous structures, the green solid curve indicates that its energy is lower than that of the line-slip structure in the same regime of $(\rho d_{\text{sphere}})^{-1}$. The shaded area corresponds to the region of all possible sphere packings which have a single value of distance R from the central axis. By definition this precludes binary mixtures. The inset shows examples of a homogeneous and a line-slip structure. See the text for an interpretation of the two red arrows in the main figure.

Intervening between these homogeneous uniform packings at specific values of D are the so-called line-slip structures. A line-slip structure corresponds to a loss of contacts along a line that separates two spiral chains. Such structures are discussed in detail in Subsection 3.3.2.

The lowest (rotational) energy per sphere (i.e., the minimal radial distance R) as a function of the dimensionless inverse number density $(\rho d_{\text{sphere}})^{-1}$ (Figure 6.4) can be obtained from established results of hard-sphere packings, as follows: For a rotational speed ω, the rotational energy per sphere is given by

$$E_{\text{rot}} = \frac{1}{2}\omega^2 (I_0 + MR^2), \tag{6.1}$$

according to the parallel-axis theorem, where $I_0 = M{d_{\text{sphere}}}^2/10$ is the moment of inertia for a sphere of mass M and diameter d_{sphere}. Since this moment of inertia is independent of the distance R from the axis, we will omit this term in the following derivations. For hard-sphere packings, the centre of each sphere is located at a distance $R = (D_{\text{cylinder}} - d_{\text{sphere}})/2$ from the cylindrical axis. The rotational energy for hard-sphere packings is thus given by

$$E^{\text{H}}_{\text{rot}} = \frac{1}{8}\omega^2 M(D_{\text{cylinder}} - d_{\text{sphere}})^2 . \tag{6.2}$$

The dimensionless inverse number density $(\rho d_{\text{sphere}})^{-1}$ is related to the diameter ratio $D \equiv D_{\text{cylinder}}/d_{\text{sphere}}$ and the packing fraction ϕ as follows: For the presence of N spheres in a finite-sized system of length L, the packing fraction ϕ is given by

$$\phi = \frac{4NV_{\text{sphere}}}{\pi D^2_{\text{cylinder}} L} = \frac{2}{3}\frac{Nd_{\text{sphere}}}{L}\frac{{d_{\text{sphere}}}^2}{D^2_{\text{cylinder}}}, \tag{6.3}$$

where V_{sphere} is the volume of a sphere. In the limit of an infinitely long cylinder ($N \to \infty$ and $L \to \infty$), the number density $\rho \equiv N/L$ is considered. Rearranging Equation (6.3), we obtain

$$(\rho d_{\text{sphere}})^{-1} = \frac{2}{3}\left(\frac{d_{\text{sphere}}}{D_{\text{cylinder}}}\right)^2 \phi^{-1} \tag{6.4}$$

for the inverse number density.

Figure 6.4 shows that the homogeneous uniform structures appear at local minima in the rotational energy per sphere, which are at particular values of $(\rho d_{\text{sphere}})^{-1}$. We therefore expect an observation of homogeneous uniform structures at these values of $(\rho d_{\text{sphere}})^{-1}$. In the intervening ranges of $(\rho d_{\text{sphere}})^{-1}$, however, binary mixtures of adjacent homogeneous structures are to be expected because their energies as obtained from the usual Maxwell (common tangent) construction[1] are lower than the energies of the corresponding line-slip structures in the same regimes of $(\rho d_{\text{sphere}})^{-1}$.

For any binary mixture, there exists an interface that separates the two homogeneous structures in the mixture. Associated with this interface is an interfacial energy. While this interfacial energy can be neglected in an energy computation of an infinitely long system via the Maxwell construction, this is not the case for finite-sized systems, which will play an important role in Section 6.4.

Note that the hard-sphere limit can be approached in two possible ways, as indicated by the red arrows in Figure 6.4: Following the horizontal red arrow, we approach the hard-sphere limit by a volumetric change (Chapter 5). This approach is equivalent to a reduction of pressure in the phase diagram of Figure 5.2. The other approach (vertical red arrow) corresponds to a variation of the rotational speed. This approach has been employed in the energy calculation of Subsection 6.3.2.

The sequence of structures in Figure 6.4 is in agreement with the findings of Lee *et al.* [Lee *et al.* (2017)]. In the following, we discuss how this sequence of structures changes in the case of soft spheres and/or of finite-sized systems.

6.3.2 Analytic Energy Calculation of Soft-Sphere Packings

For a rotating column of soft spheres, the energy per sphere is given by

$$E_{\text{rot}}^{\text{S}} = E_{\text{rot}} + E_o \,, \tag{6.5}$$

[1] Note that the Maxwell construction is again a method and terminology that we adopt from thermodynamics. Its predictions have been corroborated by simulations.

where the second term corresponds to sphere–sphere interactions. We adopt the same soft-sphere interaction between sphere i and sphere j as described in Subsection 3.2.2:

$$E_{ij} = \begin{cases} 0 & \delta_{ij} < 0 \\ \frac{1}{2}k\delta_{ij}^2 & \delta_{ij} \geq 0 \end{cases} \tag{6.6}$$

where k is the spring constant, $\vec{r}_i$ is the position of sphere i, $\vec{r}_j$ is the position of sphere j, and $\delta_{ij} \equiv d_{\text{sphere}} - |\vec{r}_i - \vec{r}_j|$ is the overlap between the spheres. The total overlap energy per sphere, E_o, can be written as

$$\begin{aligned} E_o &= \frac{1}{2}\frac{k}{N}\sum_{\substack{i,j=0 \\ i\neq j}}^{N} \delta_{ij}^2 \\ &= \frac{1}{2}k\left\langle \delta_{ij}^2 \right\rangle , \end{aligned} \tag{6.7}$$

i.e., a summation over all pairwise interactions, followed by a division by the total number of spheres in the structure. The energy per sphere is thus given by

$$\frac{E_{\text{rot}}^{\text{S}}}{M\omega^2 d_{\text{sphere}}{}^2} = \frac{1}{2}\frac{R^2}{d_{\text{sphere}}{}^2} + \frac{1}{2}\frac{k}{M\omega^2}\left\langle \left(\frac{\delta_{ij}}{d_{\text{sphere}}} \right)^2 \right\rangle . \tag{6.8}$$

In a homogeneous uniform structure, each sphere is located at a distance R from the central axis and is in contact with six neighbouring spheres. From these constraints, it follows that for a given number density ρ, the energy of a homogeneous structure can *only* be varied by a uniform radial compression, expansion, or twist, if it is to remain homogeneous (as we assume here). Thus the energy per sphere corresponds to a minimisation with respect to only two variables, namely the radial position R and the twist angle α [Mughal *et al.* (2012)]. In the case of an achiral structure, this twist angle is zero as a result of structural symmetry, so that we are only left with the variable R. An analytic expression for the energies of homogeneous achiral $(l, l, 0)$ structures (i.e., $(2, 2, 0)$, $(3, 3, 0)$, $(4, 4, 0)$, etc.) is derived in Appendix C. The minimal energy can either be obtained by a root search of the gradient or by a numerical minimisation routine for scalar functions.

(a)

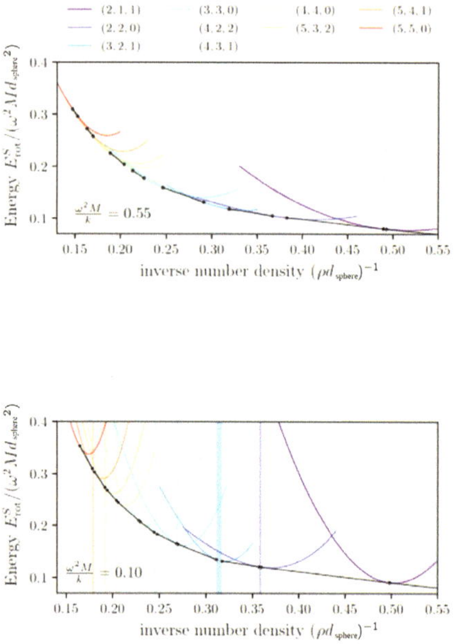

(b)

Figure 6.5 Minimal energy per sphere for all homogeneous structures as a function of dimensionless inverse number density $(\rho d_{\mathrm{sphere}})^{-1}$. The results are for soft spheres with harmonic interactions. Figures (a) and (b) correspond to the cases of $\omega^2 M/k = 0.55$ and for $\omega^2 M/k = 0.10$, respectively, where the latter is closer to the hard-sphere limit. Each black solid line denotes a common tangent between adjacent curves, each black dot denotes the contact point of a common tangent, and each curve of a unique colour denotes a particular homogeneous uniform structure. If a common tangent lies below the energy curves of homogeneous structures, a binary mixture is expected to exist. Otherwise, a homogeneous uniform structure across the system is predicted to exist, as highlighted by one of the coloured strips.

In Figure 6.5, the minimised energies of homogeneous uniform structures as a function of the inverse number density $(\rho d_{\mathrm{sphere}})^{-1}$ are displayed for two different values of $\omega^2 M/k$. These calculations are based on the assumption of harmonic inter-particle interactions between neighbouring spheres. Each black solid line denotes a common tangent between adjacent curves of homogeneous uniform structures, and each black dot denotes the contact point between a common tangent and a particular curve of homogeneous uniform structures. For any regime where a common tangent lies below the energy curves of homogeneous structures, a binary mixture is expected to exist. Otherwise, a homogeneous uniform structure is predicted to exist, as highlighted by one of the coloured strips.

The limiting case of $\omega^2 M/k \to 0$ corresponds to the hard-sphere limit [(Figure 6.5(b)]. When moving away from the hard-sphere limit (i.e., increasing $\omega^2 M/k$), the overlap between spheres increases, resulting in a broadening of parameter ranges for homogeneous structures [(Figure 6.5(a)] where there is no loss of contacts.

The corresponding phase diagram as obtained from such analytic results is presented in Figure 6.6. In the hard-sphere limit of $\omega^2 M/k \to 0$, the values of $(\rho d_{\text{sphere}})^{-1}$ for the homogeneous structures are consistent with values from the simulations of Lee *et al.* [see Figure 6.3(b) and the dashed horizontal error bars in Figure 6.6]. Lee *et al.*'s experimental data correspond to a

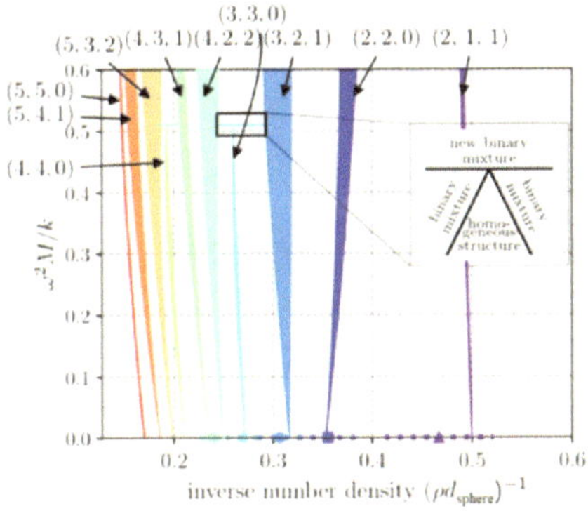

Figure 6.6 Phase diagram of homogeneous structures (coloured, labelled regions) and binary mixtures of adjacent homogeneous structures (intervening white space). Most homogeneous regions expand for increasing $\omega^2 M/k$ (i.e., rotation frequency), but the achiral $(3, 3, 0)$ and $(4, 4, 0)$ structures vanish at the corresponding *peritectoid* points (see the inset) [Abbaschian and Reed-Hill (2008)]. For the $(5, 5, 0)$ uniform structure, only the right-most boundary is shown, since this is the last homogeneous structure without inner spheres. The values of $(\rho d_{\text{sphere}})^{-1}$ for the homogeneous structures are in good agreement with values from the simulations of Lee *et al.* [Lee *et al.* (2017)], where the values of $(\rho d_{\text{sphere}})^{-1}$ from *Lee et al.* in the hard-sphere limit of $\omega^2 M/k \to 0$ are indicated by points with dashed horizontal error bars.

shift towards higher values of $(\rho d_{\text{sphere}})^{-1}$. This is possibly due to vibrations that keep the hard spheres slightly apart, yielding an effective diameter that is larger than the real diameter. A similar effect has been observed for experiments described in Section 7.5.

For increasing $\omega^2 M/k$, the regions of homogeneous structures in the phase diagram expand, as expected. The upper part of the phase diagram is, however, much richer in detail than anticipated. Of particular interest to us is the vanishing of the homogeneous achiral structures of $(3, 3, 0)$ and $(4, 4, 0)$ and their corresponding binary mixtures at a rotational velocity of $\omega^2 M/k \approx 0.5$. The points at which these achiral structures disappear are here referred to as *peritectoid* points[2] (see the inset of Figure 6.6) [Abbaschian and Reed-Hill (2008)]. At values of $\omega^2 M/k$ above each peritectoid point, the corresponding achiral structure is no longer energetically favourable when compared to their chiral counterparts because the latter can lower their own energies via a twisting deformation. A new binary mixture of the next-nearest homogeneous uniform structures emerges. The phase boundaries of the next-nearest homogeneous structures show a change in slope at the value of $\omega^2 M/k$ where the new binary mixture appears.

6.4 Columnar Structures from Rapid Rotations: Simulations of Finite-Sized Systems

The analytic results presented above are only valid for the ideal case of an infinitely long system. To corroborate and extend such investigations to finite-sized systems in reality, simulations based on the principle of energy minimisation have been carried out. The model employed in these simulations is discussed in detail in Subsection 6.4.1. Simulations of finite-sized systems have led to the unexpected discovery of a line-slip structure. In Subsection 6.4.2, we

[2] Note that a peritectoid transformation actually describes a type of isothermal reversible reaction from metallurgy, in which two solid phases react to form an alloy that is a completely different solid phase. The structures described here do not undergo any thermal reaction (since we do not have a thermal system). We still apply this term from metallurgy to the phase transitions because both share the same topological features in the phase diagram.

present the corresponding results in the form of a modified phase diagram, which is different from the phase diagram in Figure 6.6.

6.4.1 Method of Simulation: Energy Minimisation

To simulate finite-sized columnar structures assembled by rapid rotations, the soft-sphere model described in Section 5.2 has to be slightly modified. It can be used as a general simulation of a finite-sized system with N spheres in a unit cell of length L, where each sphere is allowed to occupy any position within the unit cell, but is confined by an external potential. Here, this external potential is the rotational energy E_{rot}.

This confining rotational-energy term has to be taken into account for the total energy per sphere,

$$\frac{E(\{\vec{r}_i\},\alpha)}{M\omega^2 d_{\mathrm{sphere}}{}^2} = \frac{1}{2N}\sum_i^N \left(\frac{R_i}{d_{\mathrm{sphere}}}\right)^2 + \frac{1}{2}\frac{k}{NM\omega^2}\sum_{i=1}^N\sum_{j=i}^N \frac{|\vec{r}_i - \vec{r}_j|^2}{d_{\mathrm{sphere}}{}^2}, \tag{6.9}$$

which has to be minimised with respect to all sphere positions $\vec{r}_i$ and the twist angle α at a fixed length L of the unit cell. As illustrated in Figure 6.7, we have implemented twisted periodic boundary conditions (Section 5.2) with image spheres above and below the unit cell [Mughal *et al.* (2012)].

The basin-hopping method [Wales and Doye (1997)] has been employed to search for the global minimum of energy for specific values of $\rho = N/L$ and $\omega^2 M/k$. This method is introduced in Subsection 3.5.2 and described in detail in Appendix B. Structures at $2 < \rho d_{\mathrm{sphere}} < 3$, or equivalently $0.3 < (\rho d_{\mathrm{sphere}})^{-1} < 0.5$, were studied. The number density ρd_{sphere} was varied by changing the length L of the unit cell at a fixed value of N. According to existing results for hard-sphere packings, the only possible structures in this parameter regime are the homogeneous uniform structures of $(2, 1, 1)$, $(2, 2, 0)$ and $(3, 2, 1)$ as well as their corresponding line-slip structures. To be compatible with these structures, the value of N must be a multiple of 12. In these simulations, a value of $N = 24$ was adopted.

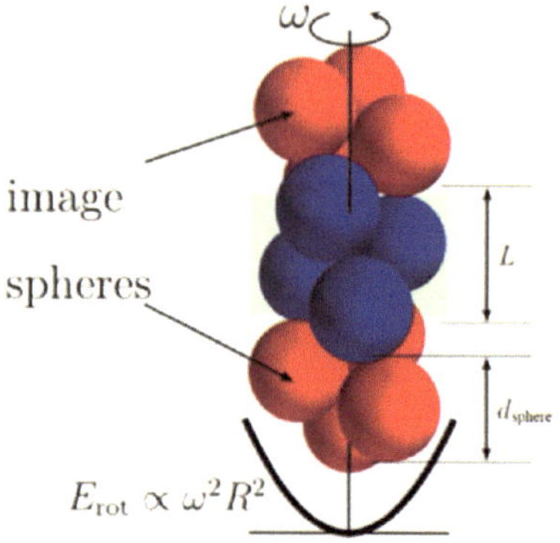

Figure 6.7 Simulation of N soft spheres (blue) of diameter d_{sphere}, confined by a rotational energy E_{rot}. Each sphere is allowed to occupy any position within the unit cell of fixed length L. The rotational energy E_{rot} is proportional to ω^2 and R^2, where ω is the rotational speed and R is the radial distance from the central axis. Twisted periodic boundary conditions, with image spheres above and below the unit cell, are employed.

6.4.2 Line-Slip Structures of Finite-Sized Systems

We compare the analytic results for the energy of an infinitely long system with the numerical results for the energy of a finite-sized system. An example at a relatively low rotational velocity (i.e., at $\omega^2 M/k = 0.2$) is presented in Figure 6.8. For the energy of an infinitely long system, the blue and the brown solid curve correspond to analytic results of the structures of $(3, 2, 1)$ and $(2, 2, 0)$, respectively. The energy of the binary mixture is given by the solid black line that denotes the common tangent between the blue and the brown solid curve. The red dotted curve, which corresponds to the numerical results for the energy of a finite-sized system, follows the trend of the analytic lowest-energy curve in each regime. However, a notable difference arises for the regime of $0.3205 \leq (\rho d_{\text{sphere}})^{-1} \leq 0.3520$. As a result of finite-size effects, the energy of a finite-sized system is slightly higher than that of an infinitely long system, as indicated by the discrepancies between the red dotted curve and the solid black curve (common tangent). In between $(\rho d_{\text{sphere}})^{-1} = 0.3205$ (vertical blue dashed line) and $(\rho d_{\text{sphere}})^{-1} = 0.3333$ (vertical red dashed line), simulations of a finite-sized system

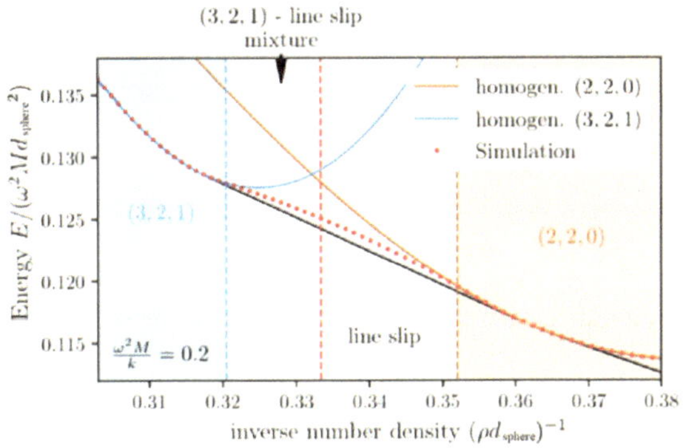

Figure 6.8 The red dotted curve is an energy curve obtained from simulations of a finite-sized system at $\omega^2 M/k = 0.2$ and $N = 24$. For the energy of an infinitely long system, the blue solid curve and the brown solid curve correspond to analytic energy curves of the uniform structures of $(3, 2, 1)$ and $(2, 2, 0)$, respectively, and for binary mixtures the solid black curve denotes the common tangent of these two curves. In between the vertical blue dashed line and the vertical red dashed line, simulations of a finite-sized system reveal the presence of a mixture of the $(3, 2, 1)$ uniform structure and the $(2, \mathbf{2}, 0)$ line-slip structure. And in between the vertical red dashed line and the brown vertical dashed line, only the $(2, \mathbf{2}, 0)$ line-slip structure has been observed. In both of these regimes, the energy of a finite-sized system is slightly higher than that of an infinitely long system, as indicated by the discrepancies between the red dotted curve and the solid black curve (common tangent).

reveal the presence of a mixture of the $(3, 2, 1)$ uniform structure and the $(2, \mathbf{2}, 0)$ line-slip structure. In between $(\rho d_{\text{sphere}})^{-1} = 0.3333$ (vertical red dashed line) and $(\rho d_{\text{sphere}})^{-1} = 0.3520$ (brown vertical dashed line), only the $(2, \mathbf{2}, 0)$ line-slip structure has been observed in simulations.

In part, these results corroborate those of the analytic treatment—in particular in regards to the homogeneous phases—but the intervention of the line slip was an unexpected effect of finite size. In these finite-sized systems the interfacial energy of the binary mixtures has a finite contribution to the total energy, which decreases with increasing number of spheres. The binary mixture of two homogeneous phases for the range of number densities presented here, can already be recovered by using system sizes of

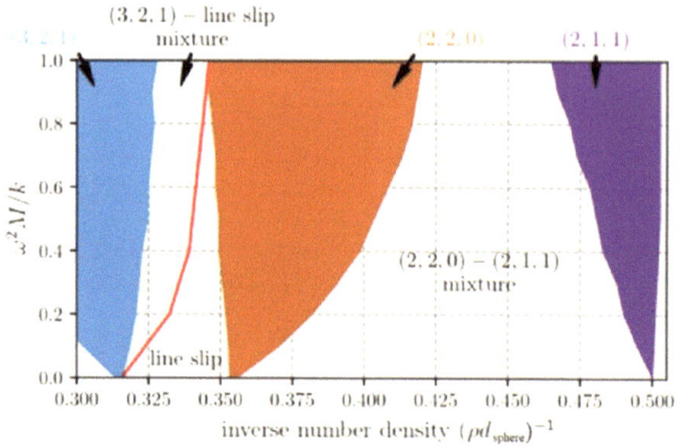

Figure 6.9 Phase diagram obtained from finite-sized simulations with $N = 24$ spheres. The intervening region between the uniform structures of $(2, 2, 0)$ and $(2, 1, 1)$ is occupied by $(2, 2, 0) - (2, 1, 1)$ binary mixtures, but the intervening region between the uniform structures of $(3, 2, 1)$ and $(2, 2, 0)$ is partly occupied by an unexpected line-slip structure.

$N = 48$ and $N = 96$. The energy per sphere for these system sizes approaches that of the common tangent.

For a finite-sized system with $N = 24$ spheres, a phase diagram (Figure 6.9) has been obtained from such simulation results. The intervening region between the uniform structures of $(2, 2, 0)$ and $(2, 1, 1)$ is occupied by a $(2, 2, 0) - (2, 1, 1)$ binary mixture as expected from the analytic results. However, the intervening region between the uniform structures of $(3, 2, 1)$ and $(2, 2, 0)$ is split into two parts, featuring the line-slip structure mentioned above at low values of $\omega^2 M/k$.

6.5 Conclusions

On the basis of a generic soft-sphere model, the phase diagram in Figure 6.6 provides a theoretical guide to the expected columnar structures for a sufficiently long rotating column of spheres. This phase diagram exhibits some interesting features, for example, the two peritectoid points above which the corresponding achiral structures are expected to vanish.

In part the simulation results corroborate those of the analytic treatment—in particular as regards to the homogeneous phases—but the intervention of the line slip was an unexpected finite-size effect. It is to be expected that line slips will play a role in all the other parts of the phase diagram, in simulations of finite-sized systems.

Future work should include an exploration of the wide range of other unexpected structures as well as a rigorous analysis of the asymptotic trend at $N \rightarrow \infty$. Results for the line slip investigated here indicate that the binary mixture of two homogeneous phases is recovered in that limit. There also remains the case of hard-wall boundary conditions at both ends of the tube in a finite-sized system, where this case is more directly relevant to the present experiments. The simulations discussed in this chapter can easily be modified in this direction by replacing the periodic boundaries with a wall potential at both ends.

It would be worthwhile to conduct further investigations into the properties of peritectoid points by simulations as well as experiments. In particular, it would be worthwhile to find out whether these peritectoid points also exist for finite-sized systems. Using the simulation methods introduced in this chapter, their relevance to finite-sized systems may be studied. Possible experiments in this direction may be performed using hydrogel spheres. (This is discussed in greater details in Chapter 8.)

The finite-size effects of line-slip structures should also be investigated in future experiments. Since we expect such finite-size effects to be present in the hard-sphere limit as well, such experiments can conveniently be done with polymeric beads, as already used in experiments by Lee *et al.* In any such experiments, one should also be aware of the existence of metastability and hysteresis in macroscopic systems, as discussed in Section 5.4 in a related context. All possible future investigations to this topic using simulations as well as experiments will be discussed in detail in Chapter 8.

Chapter 7

Hard-Sphere Chains in a Cylindrical Harmonic Potential

Relevant journal publications by J. Winkelmann:

- J. Winkelmann, A. Mughal, D. Weaire and S. Hutzler, EPL 127, 44002 (2019)

A linear chain of hard spheres confined by a cylindrical harmonic potential, with hard walls at its ends, exhibits a variety of buckled structures as it is compressed longitudinally. The structures obtained as such generally depend on the inter-particle interactions, the confining potential, and the boundary conditions. In this chapter we show that these may be conveniently observed using the assembly method of Lee et al. *(Section 6.2). If the length of the tube is commensurate with the number of spheres, the spheres self-assemble into a linear chain. For a shorter tube, the system is compressed along the axial direction and the chain will exhibit a localized buckling. We also present a corresponding theoretical model that is easy to investigate numerically as well as by analytic approximations. A wide range of predicted structures emerge via bifurcations, where the stable*

Columnar Structures of Spheres: Fundamentals and Applications
Jens Winkelmann and Ho-Kei Chan

ISBN 978-981-4669-48-1 (Hardcover), 978-0-429-09211-4 (eBook)
www.jennystanford.com

ones have also been observed in experiments. Qualitatively similar structures have also previously been found in trapped ion systems.

7.1 Sphere Chains in a Cylindrical Harmonic Potential

Let us start this chapter by first discussing in detail the phenomenon of buckling of hard sphere chains and how they are related to columnar structures. In Subsection 7.1.1 we will introduce some relevant experimental examples [Landa *et al.* (2013)]. These include experiments using ion traps where ions are held in place by an optical tweezer. Structures similar to the ones described throughout this chapter have been observed in these experiments.

Such buckled structures are described in terms of their compression. This structure-defining parameter is defined in Subsection 7.1.2, followed by an in-depth discussion of the method and results of numerical simulations.

7.1.1 Localised Buckling in Compressed Sphere Chains

Consider the presence of N identical spheres inside a rotating liquid-filled cylindrical tube of length L. At each end of the tube is the presence of a flat boundary. Since the spheres are buoyant, the spheres are driven by a centripetal force towards the axis of rotation. If the number of spheres, N, is commensurate with the length L of the tube, i.e., $L = Nd_{\text{sphere}}$, the spheres will self-assemble into a linear chain as depicted in Figure 7.1(a). If it is $L < Nd_{\text{sphere}}$, the system is compressed along the axial direction, and the chain will exhibit a localized buckling as illustrated in Figure 7.1(b). We refer to such a buckled structure as a *modulated zigzag* structure. In this chapter, we present experimental as well as theoretical findings on the buckling behaviour of such systems.

This chapter addresses the following problem: *If a chain of hard spheres in a cylindrical harmonic potential is compressed between two plates along the axial direction, what would be the corresponding force-balanced configuration?* The set-up of Lee *et al.*'s lathe experiments serves as a good platform for such investigations.

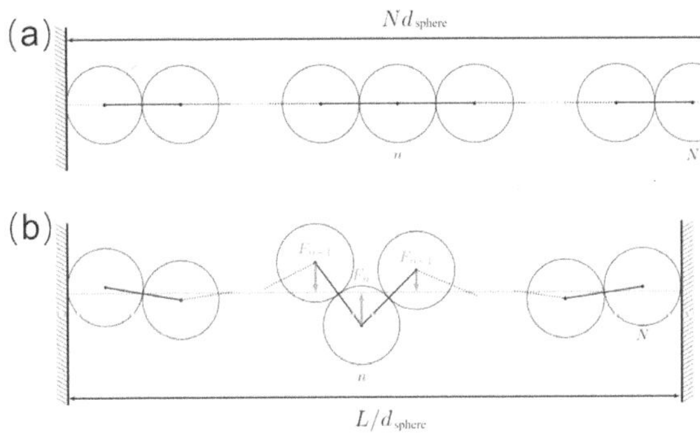

Figure 7.1 (Colour online) A chain of N spheres of diameter d_{sphere} inside a cylindrical harmonic potential and confined between two planar-wall boundaries that are separated apart by a distance L. Figure (a) shows an uncompressed linear chain at $L = Nd_{\text{sphere}}$. At $L < Nd_{\text{sphere}}$, the system is compressed along the axial direction and the chain will exhibit a localized buckling. As illustrated in Figure (b), due to the cylindrical harmonic potential, each sphere is pushed by a force F_n towards the axis of rotation (dashed line).

Previous experiments on this problem involved the use of ion traps [Mielenz *et al.* (2013); Pyka *et al.* (2013); Partner *et al.* (2015); Thompson (2015); Nigmatullin *et al.* (2016); Yan *et al.* (2016)], colloids [Hunt *et al.* (2010); Straube *et al.* (2013)], finite dust clusters [Melzer (2006)], overdamped colloids systems [Straube *et al.* (2011)] or microfluidic crystals [Beatus *et al.* (2006)]. [Landa *et al.* (2013)]. In the experiments by Landa *et al.*, a total of N ions were trapped within a quadripolar confining potential, where the corresponding ion–ion interactions are described by Coulomb's law. The traps were realised experimentally by an oscillating electric field at some trapping frequency ω. The complex variety of observed zigzag structures was illustrated by Landa *et al.* in the form of a bifurcation diagram (Figure 7.2). For ion traps that hold the ions in place, the trapping frequencies of laser along the axial direction and the radial direction are denoted by ω_x and ω_y, respectively. The parameter $\gamma_y \equiv \omega_y^2/\omega_x^2$ is a measure of the strength of cylindrical confinement. The larger the value of γ_y, the stronger the corresponding cylindrical confinement. For large values of γ_y, only the structure of a linear chain shows up. At some smaller values of γ_y, the linear chain becomes unstable and two equally stable (i.e., degenerate) zigzag structures emerge.

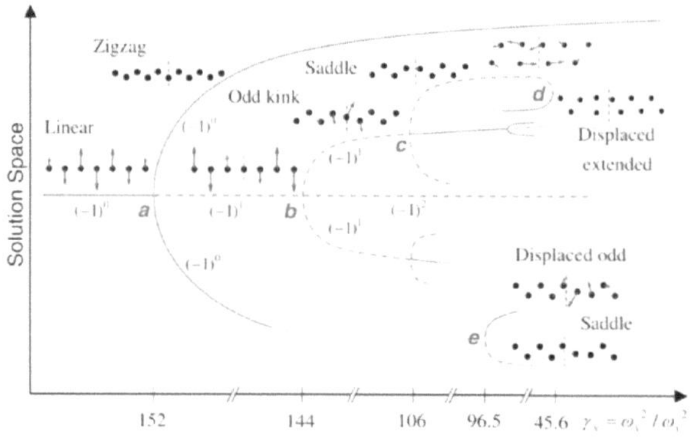

Figure 7.2 Schematic bifurcation diagram of a variety of zigzag structures as observed in experiments with ion traps [Landa *et al.* (2013)]. For ion traps that hold the ions in place, the trapping frequencies of laser along the axial direction and the radial direction are denoted by ω_x and ω_y, respectively. The parameter $\gamma_y \equiv \omega_y^2/\omega_x^2$, which decreases from left to right along the horizontal axis, is a measure of the strength of cylindrical confinement. The solid curves correspond to stable configurations, and the dashed lines correspond to unstable solutions with at least one eigenvalue of the Hessian matrix being negative. For each structure, the spatial configurations of ions and the corresponding line of axial symmetry are illustrated in the diagram.

The stability of such structures can be determined using the Hessian matrix. The presence of one or more negative eigenvalues in the Hessian indicates that a structure is unstable. In Figure 7.2 the number of negative eigenvalues, n, is displayed in the form of $(-1)^n$. At each bifurcation point, the sum of local indices from all solutions is conserved before and after the bifurcation.

Bifurcation theory can be employed to study changes in topological structures of differential-equation solutions. A bifurcation occurs when a small change in the bifurcation parameter causes an abrupt topological change. For the systems considered in this chapter, abrupt topological changes in the modulated zigzag structures occur upon a small change in the corresponding bifurcation parameter. This parameter is the strength of cylindrical confinement, γ_y, in the case of ion traps, and it is the compression Δ (the definition of Δ is presented in Subsection 7.1.2) in the case of rapid rotations. The appearance of bifurcation suggests that such a system may be described by a set of differential equations.

7.1.2 Compression Δ

In Chapter 6, the number density ρ was chosen to be the structure-defining parameter for hard-sphere structures in infinitely long systems. In finite-sized systems without periodic boundaries, ρ is not a suitable parameter since the number density varies locally.

Instead, we choose the compression

$$\Delta \equiv \frac{Nd_{\text{sphere}} - L}{d_{\text{sphere}}} = N - L/d_{\text{sphere}}, \tag{7.1}$$

to be the structure-defining parameter, where d_{sphere} is the spheres' diameter and L is the length of the confining tube. This dimensionless parameter is a measure of how much of L is reduced upon compression, as compared to the length Nd_{sphere} of an uncompressed linear chain.

This chapter concerns modulated zigzag structures at $\Delta < 1.3$, where the zigzag structures are all *planar*. Following the discussions in Chapter 6, this corresponds to a number density close to unity. Hence, the structures investigated in this chapter also appear in the phase diagram of Figure 6.6, but outside the displayed range.

7.2 Simulation Methods

Stationary structures observed within the above-mentioned regime of Δ can be predicted by two numerical methods, namely (i) an iterative stepwise method for the force-balanced positions of particles in a planar geometry, and (ii) a simulation method based on the principle of energy minimisation. While (i) restricts the hard-sphere structures within a plane, the simulation method of (ii) allows all spheres to assemble in 3D space.

For any modulated zigzag structure, both numerical approaches are only concerned with the equilibrated structures of spheres, and neglect the detailed movements of individual spheres during the assembly process.

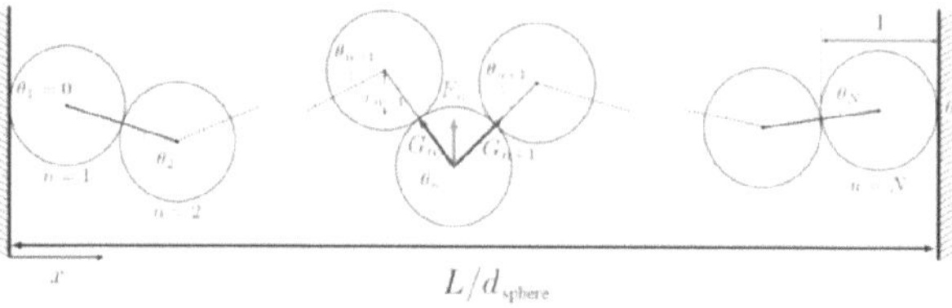

Figure 7.3 Configurations of spheres near each compressing wall and in the interior, for a modulated zigzag structure of N spheres. Sphere n, which is displaced from the central axis (dashed line) by a dimensionless distance r_n, experiences a dimensionless centripetal force $F_n = r_n$ that pulls the sphere towards the central axis. The inter-particle compressive force G_n acting on sphere n subtends an angle θ_n with the central axis. At the compressing walls, we have $\theta_1 = \theta_{N+1} = 0$.

7.2.1 Iterative Stepwise Method

A sphere of mass m inside a rotating liquid-filled tube experiences a centripetal force $f_c = m\omega^2 R$, where R is the distance of the sphere's centre from the central axis of the tube and ω is the rotational speed. For cases of zero compression ($\Delta = 0$), the spheres align in the form of a linear chain along the central axis. At a finite compression ($\Delta > 0$), the chain starts to buckle and the force-balanced structure is a modulated zigzag structure (Figure 7.3).

For each sphere, consider the dimensionless distance $r = R/d_{\text{sphere}}$ from the central axis, and the dimensionless centripetal force

$$F = f_c/(m\omega^2 d_{\text{sphere}}) = r. \tag{7.2}$$

Our aim is to calculate the dimensionless forces $F_n = r_n$ and tilt angles θ_n, as illustrated in Figure 7.3, for sphere $n = 1$ to sphere $n = N$.

Let G_n be the compressive force between sphere $(n-1)$ and sphere n. For sphere n, the condition of force balance along the axial direction is given by

$$G_n \cos\theta_n = G_0, \tag{7.3}$$

where G_0 is the magnitude of the compressive force at each end of the system. The condition of force balance along the perpendicular direction with respect to the axial direction is given by

$$\begin{aligned} F_n &= G_n \sin\theta_n + G_{n+1} \sin\theta_{n+1} \\ &= G_0(\tan\theta_n + \tan\theta_{n+1}). \end{aligned} \tag{7.4}$$

Since the centre-to-centre separation between two equal-sized spheres in mutual contact is equal to their diameter d_{sphere}, we have

$$\begin{aligned} r_n + r_{n+1} &= \sin\theta_{n+1}\,, \\ F_n + F_{n+1} &= \sin\theta_{n+1}\,. \end{aligned} \tag{7.5}$$

It follows that θ_{n+1} and F_{n+1} are related to θ_n and F_n as

$$\begin{aligned} \theta_{n+1} &= \arctan\left(\frac{F_n}{G_0} - \tan\theta_n\right), \\ F_{n+1} &\doteq \sin\left[\arctan\left(\frac{F_n}{G_0} - \tan\theta_n\right)\right] - F_n. \end{aligned} \tag{7.6}$$

For any given value of G_0, solutions to the above equations can be obtained via a "shooting method". The hard-wall boundary condition for sphere $n = 1$ requires the first tilt-angle θ_1 to be zero, regardless of the value of F_1. Starting from some arbitrarily chosen value of F_1 and using Equations (7.6), we proceed iteratively to (F_{N+1}, θ_{N+1}). The angle θ_{N+1} corresponds to the contact of the Nth sphere with the wall, as illustrated in Figure 7.3.

We search for values of F_1 (in general more than one) such that the angle θ_{N+1} is zero, satisfying the boundary condition at the other hard wall. This search is performed by "coarse-graining" the value of F_1 over a range of $0 < F_1 \leq 0.01$ in steps of 10^{-4}. These values are then used as brackets in a search via the bisection method. This root-searching method is employed for different values of G_0.

With the tilt angles of all spheres determined, the compression Δ from Equation (7.1) is given by

$$\Delta = N - \sum_{n=1}^{N} \cos\theta_n\,. \tag{7.7}$$

The dimensionless total energy E of such a hard-sphere packing consists only of the rotational energy,

$$E = \frac{E_{\text{rot}}}{m\omega^2 d_{\text{sphere}}{}^2} = \frac{1}{2}\sum_{n=1}^{N} r_n^2, \tag{7.8}$$

where, as in the case of Chapter 6, we have chosen to omit the constant contribution from the moment of inertia of each sphere.

7.2.2 Simulations Based on Energy Minimisation

To verify results obtained from the iterative stepwise method, the corresponding force-balanced configurations as computed by minimising the energy of some initial random configurations are presented here as well. Such simulations, which do *not* restrict the structures to be planar, help us find out whether the structures obtained from the iterative stepwise method are lowest-energy configurations.

For a system of $N = 20$ particles, the iterative stepwise method does not work well beyond a compression of $\Delta \geq 0.9$, where the bisection method fails to yield all the solutions. Solutions in this regime can instead be obtained by minimisation. Results for *soft*-sphere packings can be obtained first and then extrapolated to the hard-sphere limit. The methodology of such simulations is discussed in Chapter 5 and Chapter 6.

For a system of N *soft* spheres confined within a length L along the axial direction, the dimensionless total energy is given by

$$E^{\text{S}} = \frac{1}{2}\sum_{n=1}^{n=N} r_n^2 + \frac{1}{2}\left(\frac{k}{m\omega^2}\right)\left\{\sum_{\substack{n,m=1\\ m<n}}^{N}\left(\frac{\delta_{nm}}{d_{\text{sphere}}}\right)^2 + \left[\left(\frac{\delta_1}{d_{\text{sphere}}}\right)^2 + \left(\frac{\delta_N}{d_{\text{sphere}}}\right)^2\right]\right\}. \tag{7.9}$$

The very first term of the expression corresponds to the rotational energy of each sphere. The first term inside the curly brackets corresponds to inter-sphere overlaps. The overlap between sphere

n and sphere m is defined as $\delta_{nm} \equiv d_{\text{sphere}} - |\vec{R}_n - \vec{R}_m|$, where $\vec{R}_n$ and $\vec{R}_m$ are respectively the central positions of sphere n and sphere m. The second term inside the curly brackets corresponds to sphere-wall overlaps at the two ends of the tube, where δ_1 and δ_N describe this type of overlap for sphere 1 and sphere N, respectively. Both energy terms inside the curly brackets only contribute to the total energy if the overlap is positive, i.e., $\delta_{nm} > 0$, $\delta_1 > 0$, or $\delta_N > 0$.

For any given values of Δ and $k/m\omega^2$, a stable or metastable solution is obtained by adjusting the central positions of spheres to achieve an energy minimum. A local-minimisation routine, such as the gradient-descent method and BFGS method (as discussed in Subsection 3.5.1 and Appendix B), is suitable for such simulations. By performing a series of simulations for increasing $k/m\omega^2$, the soft-sphere results can be extrapolated to the hard-sphere limit of $k/m\omega^2 \to \infty$.

Solutions from the iterative stepwise method are obtained from conditions of force balance, where such solutions can be stable or unstable. The stability of a solution can be checked by subjecting the corresponding structure to a random perturbation and then by applying the steepest-descent algorithm to determine whether the structure would return to its initial configuration.

7.3 Numerical Results

In this chapter, we mainly discuss structures that exist in the low-energy and small-compression regimes. As hinted in Figure 7.5, there also exist a rich variety of structures in the high-energy and/or large-compression regimes, but those structures are outside the scope of our discussion.

7.3.1 Typical Profiles

For cases of a very small compression, only the *symmetric* structure S, where the profiles of F_n and r_n are symmetric about the midpoint of the system, has been obtained from the iterative stepwise method. Note that the iterative stepwise method yields only absolute values for F_n and r_n. For cases of a relatively large compression,

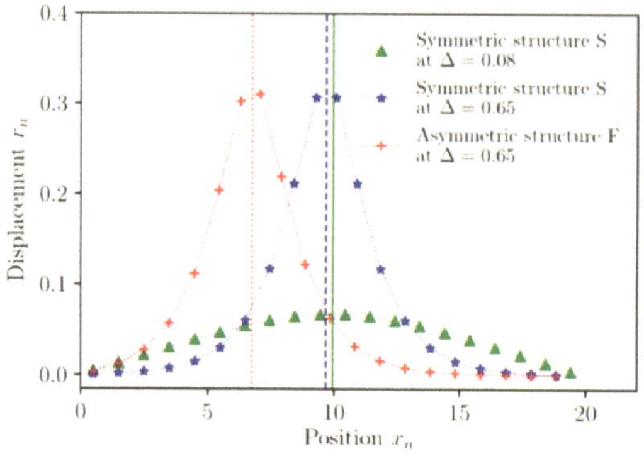

Figure 7.4 Dimensionless displacement r_n of the hard spheres from the central axis as a function of the dimensionless position x_n, for a symmetric structure (blue stars) and an asymmetric structure (red crosses) at a compression of $\Delta = 0.65$. Also shown in the figure is the displacement profile of a symmetric structure (green triangles) at a smaller compression of $\Delta = 0.08$. The vertical blue dashed line and the green solid line indicate the midpoints for the cases of $\Delta = 0.65$ and $\Delta = 0.08$, respectively. The peak position of the asymmetric structure at $\Delta = 0.65$, as estimated by a quadratic fit of the displacements around the peak, is indicated by the vertical dotted red line. The distance between x_0 and x_N is equal to $(N - \Delta - 1)$.

both symmetric and asymmetric structures have been obtained. In Figure 7.4, the dimensionless displacement r_n from the central axis is plotted against the dimensionless position

$$x_n \equiv 1/2 + \sum_{i=2}^{n} \cos(\theta_i) \tag{7.10}$$

for stable structures of $N = 20$ spheres at $\Delta = 0.08$ (very small compressions) and $\Delta = 0.65$ (relatively large compressions), respectively. At $\Delta = 0.08$, only a symmetric structure S exists, as indicated by the green triangles. The dimensionless displacements for this structure is in good agreement with those of the symmetric structure generated by energy-minimisation simulations. At $\Delta = 0.65$, there is the presence of a symmetric structure S (blue stars) and an asymmetric structure F (red crosses).

7.3.2 Bifurcation Diagrams

We have employed the iterative stepwise method to search for structures in the parameter ranges of $0.19 \leq G_0 \leq 0.25$ and $0 < F_1 < 0.01$. Structures at $\Delta < 0.9$ have been obtained for both an even ($N = 20$) and an odd ($N = 19$) number of spheres, where the results differ qualitatively for the two values of N. Bifurcation diagrams, with G_0 and F_1 being the corresponding variables, are presented in Figure 7.5(a) and in Figure 7.5(b) for the cases of $N = 20$ and $N = 19$, respectively. Each curve corresponds to a structure

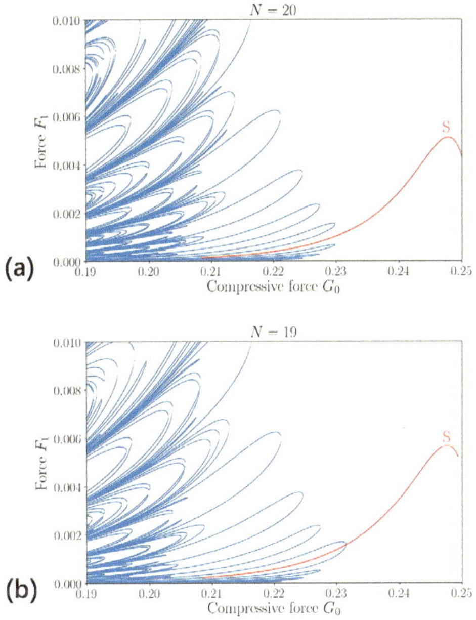

Figure 7.5 Bifurcation diagrams of F_1 against G_0 for the cases of (a) $N = 20$ and (b) $N = 19$. For each case of N, the force F_1 of the first sphere in the iterative stepwise method is plotted against the compressive force G_0. Each curve corresponds to a structure obtained by the iterative stepwise method. The red curve corresponds to the symmetric structure S, where this curve for symmetric structures cannot be resolved at $G_0 < 0.21$ as a result of the particular implementation of the iterative stepwise algorithm (Subsection 7.2.2). An asymmetric structure indicated below the red curve is related to an asymmetric structure indicated above the red curve via a mirror-symmetry operation about the midpoint of the system.

obtained by the iterative stepwise method. At large values of G_0, only the symmetric structure S (red curve) is observed. This curve for symmetric structures cannot be resolved at $G_0 < 0.21$ as a result of the particular implementation of the iterative stepwise algorithm (Subsection 7.2.2). At some smaller values of G_0, curves of asymmetric structures emerge both above and below the curve for symmetric structures. Any asymmetric structure indicated below the red curve is related to an asymmetric structure indicated above the red curve via a mirror-symmetry operation about the midpoint of the system.

These two bifurcation diagrams show that a wide range of structures can be obtained via the iterative stepwise method. However, the diagrams in terms of F_1 and G_0 do not provide any physical insights into how various structures come into existence. We have therefore turned to the use of energy diagrams, with an analysis that focuses on low-energy structures.

For low-energy structures, we have computed the relative energy

$$\Delta E = E - E_{\text{Symm}} \tag{7.11}$$

as well as their compression Δ for the cases of $N = 20$ (even number) [Figure 7.6(a)] and $N = 19$ (odd number) [Figure 7.6(b)] at a compressive force G_0. The parameter range is chosen such that the emergence of the first few asymmetric structures via bifurcations is included.

The results for the two cases (even and odd N) are qualitatively different. We therefore discuss them separately as follows: In the case of $N = 20$ [Figure 7.6(a)], the asymmetric structures A*, B, C*, D, E*, and F emerge through bifurcations for increasing Δ. Each unstable structure is marked with an asterisk. The first two asymmetric structures A* and B emerge from an "out-of-the-blue" bifurcation at $\Delta = 0.558$ without any preceding structure. Of these two structures, structure B is stable but structure A* is unstable. At $\Delta = 0.588$, structures C* and D emerge via a pitch-fork bifurcation out of structure B. At $\Delta = 0.622$, a similar pitch-fork bifurcation of structure D branch occurs for the emergence of structures E* and F.

Compared to the case of $N = 20$, the energy diagram for the case of $N = 19$ [Figure 7.6(b)] differs by the structures involved in the first bifurcation. Structure B emerges as the only structure. Subsequent

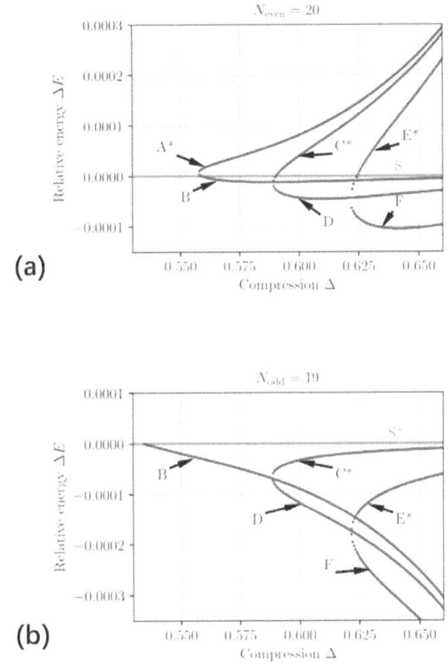

Figure 7.6 Energy-bifurcation diagrams in the vicinity of the first bifurcation: The relative energy $\Delta E = E - E_{\text{Symm}}$ is plotted against the compression Δ, for the cases of (a) $N = 20$ (even number) and (b) $N = 19$ (odd number). Each unstable structure is marked with an asterisk. Some examples of structures for the case of $N = 20$ are displayed in Figure 7.7.

bifurcations in the case of $N = 19$ follow the same pattern as their counterparts in the case of $N = 20$.

For the case of $N = 20$, Figure 7.7 displays examples that correspond to the buckled-chain structures S, A, B, C, D, E and F as labelled in Figure 7.6(a). For each example, the vertical black solid line indicates the centre of the structure. The vertical red dashed line indicates the peak position of the displacement profile. This peak is estimated via a quadratic fit to the positions of neighbouring spheres. For each unstable structure, the peak position coincides with the centre of a sphere. Note the degeneracy of asymmetric structures: For structures of the same energy, the peak may be located left or right of the midpoint of the structure.

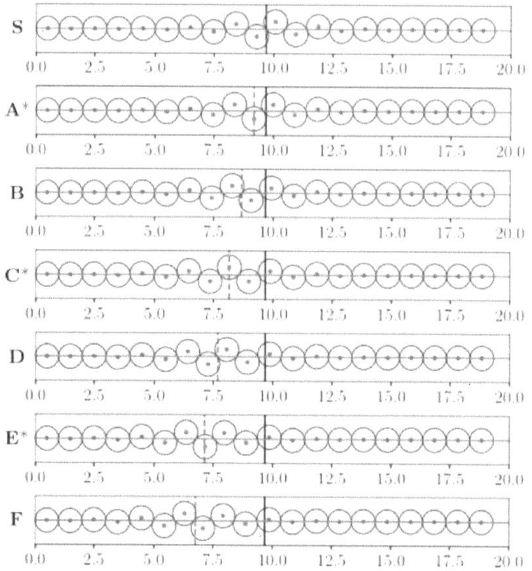

Figure 7.7 (Colour online) Examples for the buckled-chain structures S, A*, B, C*, D, E* and F as labelled in Figure 7.6(a). For the case of $N = 20$, the structures were obtained at a compression of $\Delta = 0.65$. Each unstable structure is marked by an asterisk. For each example, the solid black vertical line indicates the centre of the structure, and the vertical red dashed line indicates the peak position of the displacement profile.

7.3.3 Maximum Angles

For the symmetric structure S, we have also computed the maximum angle $\theta_{\max}$ as a function of the compression Δ. As discussed in Section 7.5, this is a quantity that can readily be extracted from experimental data. The results are presented in Figure 7.8. For cases of a small compression ($\Delta \lessapprox 0.1$), the displacement profile is of a parabolic shape [green triangles in Figure 7.4]. For cases of a larger compression, the profile becomes hyperbolic in shape [blue stars in Figure 7.4].

While results obtained from the iterative stepwise method are only up to $\Delta \sim 0.9$, the maximum angle $\theta_{\max}$ as obtained from energy-minimisation routines has been computed up to $\Delta \sim 1.3$. At this value of Δ, there is a qualitative change in the structure since each sphere acquires a new contact with its next-nearest neighbour.

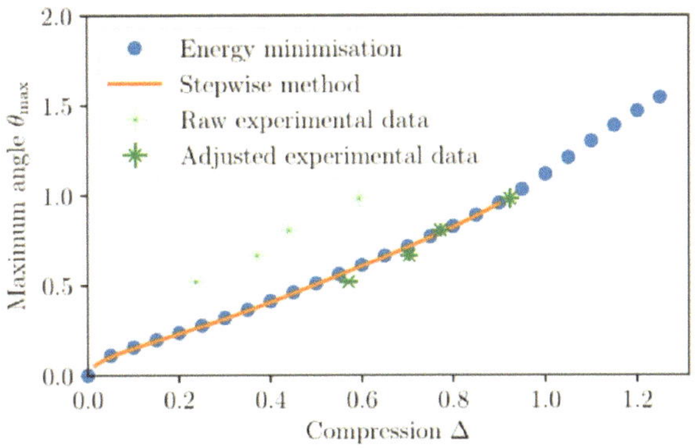

Figure 7.8 Maximum angle $\theta_{\max}$ of the symmetric structure as a function of the compression Δ, for the value of N being an even number. The blue circles and the orange solid curve correspond respectively to numerical results from the iterative stepwise method and from energy minimisation. The light-green crosses correspond to raw experimental data, and the dark-green crosses correspond to adjusted experimental data that take into account an effective diameter of spheres. This effective diameter is a result of vibrations in the experimental set-up. The uncertainty in $\theta_{\max}$ was determined by averaging the angles over five images of the structure at the same value of Δ.

7.4 Linear Approximation

To provide physical insights into the numerical results presented in the previous sections, we present a theoretical description based on a linear approximation. For sufficiently small values of θ_n and F_n, a linearisation of Equation (7.6) leads to

$$\begin{pmatrix} F_n \\ \theta_n \end{pmatrix} = \begin{pmatrix} \frac{1}{G_0} - 1 & -1 \\ \frac{1}{G_0} & -1 \end{pmatrix} \begin{pmatrix} F_{n-1} \\ \theta_{n-1} \end{pmatrix}. \tag{7.12}$$

A recursive substitution of F_n and θ_n, with $\frac{1}{G_0} = 4 + \epsilon$ for small positive values of ϵ, results in

$$\begin{pmatrix} F_n \\ \theta_n \end{pmatrix} = \underbrace{\begin{pmatrix} 3+\epsilon & -1 \\ 4+\epsilon & -1 \end{pmatrix}}_{\underline{\underline{M}}}^{n-1} \begin{pmatrix} F_1 \\ \theta_1 \end{pmatrix}. \tag{7.13}$$

The largest possible value for the compressive force is $G_0 = 1/4$. For the case of an infinitely long chain, this limit can be derived from equations for the uniform zigzag structure ($r_n = -r_{n-1}$, $F_n = F_{n-1}$, and $\theta_n = \theta_{n-1}$). For this structure, Equation (7.4) and Equation (7.5) can be written as

$$2F_n = \sin(\theta_n) \tag{7.14}$$

and

$$F_n = 2G_n \sin(\theta_n), \tag{7.15}$$

respectively, leading to the conclusion of $G_n = 1/4$. According to Equation (7.3), this is the largest possible value of G_0.

A solution for $(F_n, \theta_n)^{\mathrm{T}}$ can be expressed in terms of the eigenvalues $\lambda_{1,2}$ and the eigenvectors $\vec{V}_{1,2}$ of the matrix $\underline{\underline{M}}$ as

$$(F_n, \theta_n)^{\mathrm{T}} = a\lambda_1^{n-1}\vec{V}_1 + b\lambda_2^{n-1}\vec{V}_2\,. \tag{7.16}$$

The eigenvalues are given by

$$\lambda_{1,2} = 1 + \frac{\epsilon}{2} \pm \frac{1}{2}\sqrt{\epsilon^2 + 4\epsilon}\,, \tag{7.17}$$

which can be approximated as

$$\lambda_{1,2} \approx 1 \pm \sqrt{\epsilon}\,. \tag{7.18}$$

Its $(n-1)$th power can be approximated as

$$\lambda_{1,2}^{n-1} = \mathrm{e}^{\ln(1\pm\sqrt{\epsilon})^{n-1}} \approx \mathrm{e}^{\pm\sqrt{\epsilon}(n-1)}\,. \tag{7.19}$$

The corresponding eigenvectors can then be written as

$$\vec{V}_{1,2} = \left[\frac{1}{2}\left(1 \pm \frac{\sqrt{\epsilon}}{2}\right), 1\right]^{\mathrm{T}}. \tag{7.20}$$

The prefactors a and b are determined from the values of F_1 and θ_1. The solution in the linear approximation for $\theta_1 = 0$ is then given by

$$F_n = \frac{2F_1}{\sqrt{\epsilon}} \sinh[\sqrt{\epsilon}(n-1)] + F_1 \cosh[\sqrt{\epsilon}(n-1)] \tag{7.21}$$

$$\theta_n = \frac{4F_1}{\sqrt{\epsilon}} \sinh[\sqrt{\epsilon}(n-1)]\,. \tag{7.22}$$

The force F_n can be written in a more concise form as

$$F_n = \frac{F_1}{\sinh(\phi)} \sinh[\sqrt{\epsilon}(n-1) + \operatorname{arctanh}(\sqrt{\epsilon}/2)] \tag{7.23}$$

For the angle θ_n, a comparison between values computed from Equation (7.23) and values obtained from the iterative stepwise method, for $G_0 = 0.234$ or $\Delta = 0.500$, is shown in Figure 7.9. The value of F_1 in the linear approximation is the same as the corresponding value in the iterative stepwise method. There is an excellent agreement between the linear approximation and the stepwise method up to about $n = 8$ spheres. The linear approximation produces a monotonically increasing function (Figure 7.9), whereas the accurate solution "rolls over" and decreases

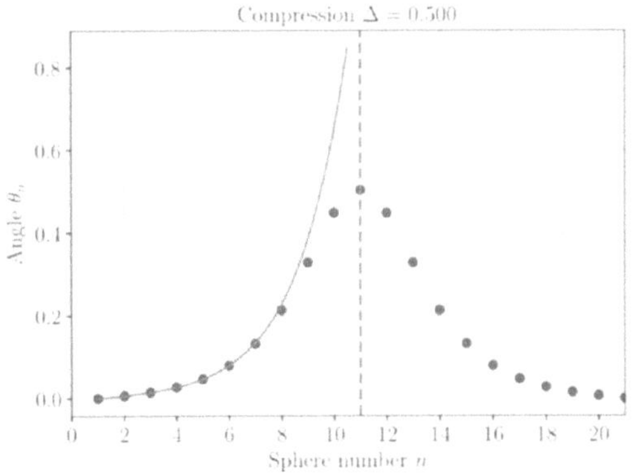

Figure 7.9 (Colour online) Variation of the angle θ_n with the sphere index n as obtained from the exact iterative stepwise method (blue dots) and from the linear approximation (blue solid line), for $N = 20$ and $\Delta = 0.500$. Note that there are 21 angles plotted for the stepwise method, since θ_{N+1} is associated with the wall contact of sphere $N = 20$.

towards the second boundary. This is due to the non-linearity of the accurate equations employed in the iterative stepwise method.

7.5 Comparison with Experiments

The experimental approach described in this section is similar to that of Lee *et al.* [Lee *et al.* (2017)]. We placed an even number of $N = 34$ polypropylene beads of density $\rho = 0.900\,\mathrm{g/cm^3}$ and diameter $d_{\mathrm{sphere}} = 3.000 \pm 0.001\,\mathrm{mm}$ [Redhill Precision Speciality Balls (Manufacturer)] inside a cylindrical tube (I.D. 15.91 ± 0.01 mm; O.D. 20.17 ± 0.01 mm; length 130.55 ± 0.01 mm) filled with water (density $\rho_{\mathrm{w}} = 1\,\mathrm{g/cm^3}$). To ensure that no air bubble can enter the tube nor stay inside the tube, each end of the tube is sealed with a stopper. The extent to which the stoppers intrude into the tube can be adjusted. As such, the value of the compression Δ can be varied. The tube is then mounted onto a commercial lathe (Charnwood W824), for which we set the rotation frequency to be $\omega = 1800 \pm 50$ rpm. A stroboscopic lamp, of which the frequency matches that of the lathe, is used to record a video of any structure formed inside the rotating system. A small off-set between both frequencies is employed, such that any recorded structure appears to be slowly rotating (see the supplemental video by Winkelmann *et al.* [Winkelmann *et al.* (2019b)]).

In Figure 7.10, the snapshot that corresponds to structure P is an in-plane view of one of the modulated zigzag structures, illustrating the planarity of experimental structures as assumed theoretically in the numerical iterative stepwise method. The other structures correspond to the same structures of S, B, D, and F as indicated in Figure 7.6 and Figure 7.7. Each structure, as identified by the corresponding distance between the peak position and the centre, corresponds to a particular stable structure indicated in Figure 7.6. Structure S and structure B were obtained at $\Delta = 0.44 \pm 0.02$, structure D was obtained at $\Delta = 0.59 \pm 0.02$, and structure F was obtained at $\Delta = 0.68 \pm 0.02$.

To reconcile experimental results with corresponding theoretical predictions, it is necessary to introduce for the spheres an *effective* diameter, which is about $1\,\%$ greater than the real value. This effectively increases the value of Δ by approximately

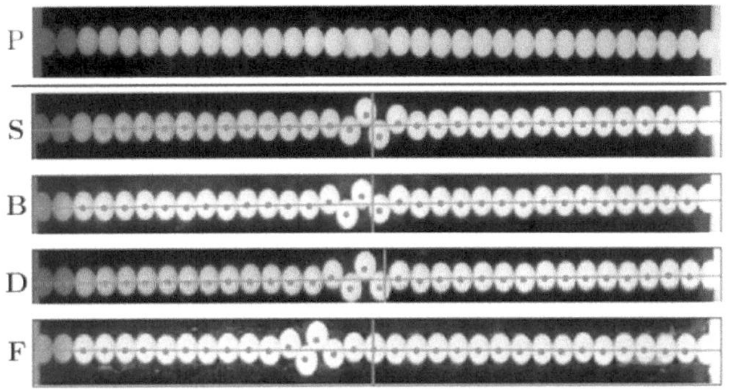

Figure 7.10 (Colour online) Snapshots of experimentally obtained buckled-chain structures for a system of $N = 34$ particles. The structures were obtained by the lathe-driven rapid rotations of a water-filled tube with the presence of polypropylene beads (density of $\rho = 0.900$ g/cm^3). The modulated zigzag structure labelled as P is rotated by $\pi/2$ with respect to the other structures. It illustrates the planarity of the experimental structures. The other structures correspond to the same structures of S, B, D, and F as indicated in Figure 7.6 and Figure 7.7, where for each of these structures the vertical green line indicates the midpoint of the system and the horizontal green line represents the central axis of rotation.

0.35, and yields a quantitative agreement between experimental and numerical results (Figure 7.8). The need to introduce such an effective diameter is possibly a result of vibrations in the lathe. As discussed in Chapter 6, this shift in Δ also appears in Lee *et al.*'s experiments. It was observed by comparing the numerical and experimental results in Subsection 6.3.2.

7.6 Conclusions

The simple system of a longitudinally compressed chain of spheres presents a rich variety of structural behaviour, as previously described in terms of "kinks" or "solitons" [Partner *et al.* (2015)]. We have explored many properties of such a system, using simple experimental apparatus and theoretical methods. We hope that such results will find applications in other types of physical systems, for example ions confined in optical traps [Landa *et al.* (2013); Mielenz *et al.* (2013); Pyka *et al.* (2013); Straube *et al.* (2013); Dessup *et al.* (2015); Partner *et al.* (2015); Thompson (2015); Nigmatullin *et al.* (2016); Yan *et al.* (2016)]. In general, these findings should be relevant to any physical system in which buckling is a key feature, for example, recent experiments that involve an expanding (growing) elastic beam pinned to a substrate [Michaels *et al.* (2019)].

While we have only observed relatively simple structures with single peaks in their displacement profiles, some more complicated structures exist in the high-energy and/or high-compression regimes, as hinted by the bifurcation diagram in Figure 7.5. Among these are structures that are created by concatenating one of the single-peak structures with its mirrored counterpart.

For the case of $N = 20$ (even number), we have shown that all the stable structures as predicted by the numerical simulations introduced in this chapter also show up in experiments. For the maximum angle θ_{max}, we have only carried out a comparison between experimental and numerical results for the symmetric structure. For future work, this comparison between experiments and simulations should be extended to other observed structures and to other values of N. Structures may also be compared in terms of their rotational energy for an extended range of values of Δ and N. The rotational energy of each experimental structure (S, B, D, and F) can be calculated from the extracted positions of spheres.

For a compressed chain of spheres, there is currently much interest in the motion of the single peak and the corresponding Peierls-Nabarro potential. This is the potential needed to move the single-peak position so as to transform one stable structure to another, for example, from structure B to structure S (Figure 7.7). It is possible to estimate such a potential by a smooth interpolation of the energy values of stable and unstable states, as illustrated in Figure 7.11. Future work should include a corresponding experimental characterisation of this potential.

Recently, a yet simpler experimental method for generating buckled-chain structures has been discovered [Hutzler *et al.* (2020)]. The experimental set-up consists of a horizontal tube, into which ball bearings are introduced. A slight agitation of the ball bearings enables them to settle in modulated zigzag structures similar to those described in this chapter.

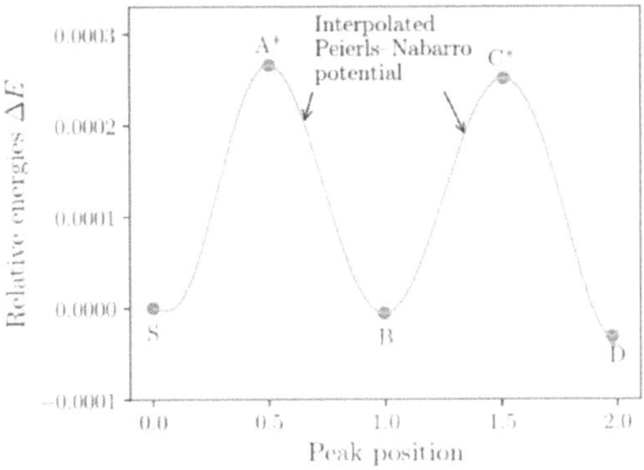

Figure 7.11 (Colour online) Illustration of the Peierls–Nabarro potential. For each of the structures S, A*, B, C* and D as indicated in Figure 7.7, each blue dot indicates the corresponding relative energy ΔE (vertical axis) and peak position (horizontal axis). The orange solid curve corresponds to a smooth interpolation through the blue dots. The stable structures S, B and D are at the minima of the interpolation, and the unstable structures A* and C* are at the maxima.

An alternative approach, which involves bubbles inside a horizontal liquid-filled tube, also yield similar buckled-chain structures. In the experiments, the bubbles are pushed towards the curved wall of the tube by buoyancy, and are confined by a potential that is, to a first approximation, quadratic. A modulated zigzag structure is formed when the bubbles are compressed along the axial direction. Other variations of the experimental method described in this chapter involve the use of soft (elastic) spheres, for example, hydrogel particles.

As to future work, it would be worthwhile to extend such investigations to the large-compression regimes, so as to make a connection with the three-dimensional structures discussed in Chapter 6. It would also be worthwhile to employ an experimental technique [Majmudar and Behringer (2005); Daniels *et al.* (2017)] that uses photoelastic materials to track the magnitude of the compressive forces. These different experimental approaches for future investigations are discussed in detail in Chapter 8.

Chapter 8

Summary and Outlook

In this chapter, we summarise the key scientific findings of columnar structures as presented from Chapter 4 to Chapter 7, and propose possible future work for simulations and experiments. Such future work can help develop potential applications in the areas of botany, foams, and nanoscience, as discussed in Section 2.4.

8.1 Chapter Outline

The focus of this monograph is on columnar structures of spheres in three-dimensional space, of which the important results will be summarised in the first part of this chapter. In the second part of this outlook, we describe the two-dimensional counterparts of columnar structures in detail, how they can be studied using soft disks, and how the corresponding results could be useful to research in microfluidics.

While the soft-sphere model, extensively used throughout this research, has shown its great advantages, there are also limitations as to its applicability to soft matter systems such as foams. These limitations are discussed in this chapter, with a comparison of results from the soft-disk model and numerical results of two-dimensional

Columnar Structures of Spheres: Fundamentals and Applications
Jens Winkelmann and Ho-Kei Chan

ISBN 978-981-4669-48-1 (Hardcover), 978-0-429-09211-4 (eBook)
www.jennystanford.com

foams. In addition, we discuss a model that is based on a theory by Morse and Witten and that might be used in future investigations of columnar structures.

8.2 Hard-Sphere Packings from Sequential Deposition

8.2.1 Summary

In Chapter 4, we showed that the densest possible columnar structures of equal-sized spheres at $D \in [1, 1 + 1/\sin(\pi/5)]$ can be constructed via a relatively simple method of sequential deposition [Chan (2011, 2013); Chan *et al.* (2019)]. This corresponds to a regime of diameter ratios D where the spheres of any densest possible columnar structure are all in contact with the cylindrical wall.

In this method, spheres that are in contact with the cylindrical wall are dropped one by one into their lowest possible positions, resulting in a local maximisation of the packing fraction. For any given value of D, the underlying template that comprises the first few spheres is fine-tuned in a search for the corresponding densest possible columnar structure.

This packing algorithm not only serves the purpose of reproducing known densest possible columnar structures, but also leads to unexpected discoveries of novel ordered columnar structures (albeit not being the densest possible ones) upon a variation of the underlying template. An example of such a novel structure is the hybrid-helix structure [Chan (2013)] shown in Figure 4.17 and Figure 4.18.

8.2.2 Outlook

The findings concerning the method of sequential deposition are largely empirical. Future work should focus on a theoretical understanding of how ordered columnar structures emerge from this deposition algorithm. The algorithm also opens up new avenues for the development of "template-controlled columnar crystallography". Future work should include an extensive search for novel ordered but non-densest structures at $D \in [1, 1 + 1/\sin(\pi/5)]$ and beyond. It would also be worthwhile to develop a similar sequential-deposition algorithm for cases of $D > 1 + 1/\sin(\pi/5)$ [Mughal *et al.* (2012); Fu *et al.* (2016, 2017)], where the corresponding densest-packed structures consist of internal spheres that are not in contact with the cylindrical wall.

8.3 Soft-Sphere Packings in Cylinders

8.3.1 Summary

Chapter 5 focusses on columnar structures of soft spheres in cylindrical confinement. The results presented in this chapter are an extension of previous findings of hard-sphere packings towards soft-sphere packings in cylindrical confinement.

For soft-sphere packings, we have reported on the experimental observation, as well as corresponding simulations, of a *line-slip* structure. The experiment was conducted with a column of bubbles under forced drainage, where a line-slip structure has clearly been observed. The simulations presented in this chapter were obtained from a minimisation of enthalpy. We have presented a phase diagram for all stable (i.e., minimal enthalpy) structures up to the diameter ratio $D = 2.71486$, where the nature of densest packings changes.

On the other hand, macroscopic systems of this kind are not confined to the ideal equilibrium states depicted in the phase diagram. We therefore present here explorations of hysteretic structural transitions of columnar structures by means of further simulations. An example of a reversible transition has been discussed in detail, where the results are presented in the form of a stability diagram. In addition, for the uniform structures in the phase diagram, we present all allowed structural transitions in the form of a directed network.

8.3.2 Outlook: An Exhaustive Investigation of Hysteresis

For future work, soft-sphere simulations based on enthalpy minimisation can be used for an exhaustive investigation of hysteresis and metastability. So far, we have only presented a stability diagram that is representative of reversible transitions. However, such a diagram will be significantly different for other types of transitions, such as irreversible transitions. A stability diagram for irreversible transitions covers cases in which an initial uniform structure U_1 is not recovered when the diameter ratio D or pressure p is back to its original value. In the hard-sphere limit, for example, the $(4, 2, 2)$ uniform structure would transform naturally into the $(5, 3, 2)$ uniform structure upon increasing D, but for decreasing D, the $(5, 3, 2)$ would undergo a transition into the $(4, 3, 1)$ structure instead of the $(4, 2, 2)$ structure (see also Figure 5.10 in Subsection 5.4.3). Thus, in general, the stability diagram for an irreversible transition involve three types of uniform structures, while the stability diagram for the transitions explained in this monograph only involved two. In the case of an irreversible transition, two of the three uniform structures may then co-exist in the same region of the stability diagram. There may also be more than one possible irreversible transition.

Research on metastability and hysteresis can also be extended to cases of structural transitions at $D > 2.71486$ [Fu *et al.* (2016)] (Section 3.4). In this regime of D, the densest possible packings of hard spheres consist of spheres that are not in touch with the cylindrical wall. The corresponding structural transitions can possibly be summarised in the form of a stability diagram. On the other hand, it would be worthwhile to find out how columnar structures with internal spheres at $D > 2.71486$ are related to those without internal spheres at $D \leq 2.71486$. This corresponds to an extension of the directed network in Figure 5.10 to $D > 2.71486$.

The work of Pittet *et al.* [Pittet *et al.* (1996)] on structural transitions in cylindrical dry foams can be extended to wet-foam structures, where a variety of methods can be employed to vary the diameter ratio D or pressure p. The general set-up for such experiments is discussed in Subsection 2.4.2. One method concerns a variation of D. The size of bubbles, and hence the diameter ratio D, can be controlled by the gas flow rate q_0 (see Figure 2.7 in Subsection 2.4.2). A wet-foam structure can be forced into a structural transition by a change in the gas flow rate. However, there is no guarantee that the foam bubbles would remain monodisperse. It might be more convenient to vary the pressure p. Boltenhagen *et al.*

[Boltenhagen *et al.* (1998)] used a piston to vary the pressure of dry foams. Structural transitions occur if a foam column is compressed or dilated in this way. Similar experiments can be performed for wet foams in forced drainage. One can also take advantage of the natural variation of pressure within a vertically aligned foam column. The liquid at the top of the column pushes down onto the bubbles below, leading to a vertical variation of pressure. Structural transitions due to pressure variations can be studied, where experimentally observed structural transitions can be compared to those predicted by the directed network of Figure 5.10. As a third option, structural transitions can also be investigated by inducing a shear stress to the system. In such experiments, the bubbles at the top and the bottom of a foam column are rotated or sheared about the vertical axis. This might induce a structural transition between a chiral and an achiral structure. Corresponding simulations based on the soft-sphere model can be performed, as well.

8.4 Rotational Columnar Structures of Soft Spheres

8.4.1 Summary

In Chapter 6, a novel method to assemble columnar structures involving rapid rotations around a central axis is analysed. Lee *et al.* [Lee *et al.* (2017)] used this method to drive spheres, which are of a lower density than the surrounding fluid, towards the central axis. This resulted in the observation of a variety of columnar structures.

For columnar structures obtained through this method, we have presented comprehensive *analytic* energy calculations that are based on the soft-sphere model. The corresponding phase diagram (Figure 6.6) displays a variety of interesting features, including two *peritectoid* points at which the achiral uniform structures of $(3, 3, 0)$ and $(4, 4, 0)$ vanish.

These analytic calculations were complemented by computationally inexpensive numerical simulations for finite-sized systems of soft spheres. The simulations partly corroborate our analytic results, and reveal the unexpected existence of a line-slip structure due to finite-size effects.

8.4.2 Outlook: Further Investigations of Finite-Size Effects

The research presented in Chapter 6 on this assembly method opens up a variety of possible experiments and finite-size simulations. For the simulations of finite-sized systems, we have only reported on the existence of one particular line-slip structure. Further line-slip structures might be present in other parts of the phase diagram (Figure 6.6). Therefore, future work should include an extended investigation of such structures.

Preliminary results for the discovered line-slip structure indicate that the binary mixture is recovered in the limit of an infinite number of spheres ($N \to \infty$). The corresponding total energy as a function of N for the binary mixture as well as the line-slip structure should be part of future investigations. The value of N, at which the binary mixture is recovered can also be of future interest. Similar investigations should also be performed for other unexpected (line-slip) structures. To achieve a better comparison between simulations and the real lathe experiments, simulations of finite-sized systems can be modified by adding walls to both ends of the unit cell. As a result of these additional walls, structures that are not predicted by the phase diagram may arise. It would be worthwhile to find out whether such wall-induced effects will diminish for increasing N. If so, the results would gives a hint for the experimental system size at which the observed structures resemble those of an infinite system. On the other hand, experiments with polymeric beads can be performed to search for unexpected line-slip structures of hard-sphere packings. The corresponding results should be compared with those from simulations of finite-sized systems with the presence of wall boundaries. Note that for such experiments hysteresis and metastability needs to be considered again.

To corroborate the findings presented here from soft-sphere simulations, in particular the occurrence of peritectoid points, Lee *et al.*'s type of experiments should also be performed with soap bubbles immersed in a surfactant-filled rotating tube. However, as a drawback, the bubbles in such experiments are prone to coarsening.

As an alternative, hydrogel spheres [Figure 8.1(a)] can be used as soft spheres for such experiments. Hydrogel spheres consist of

(a)

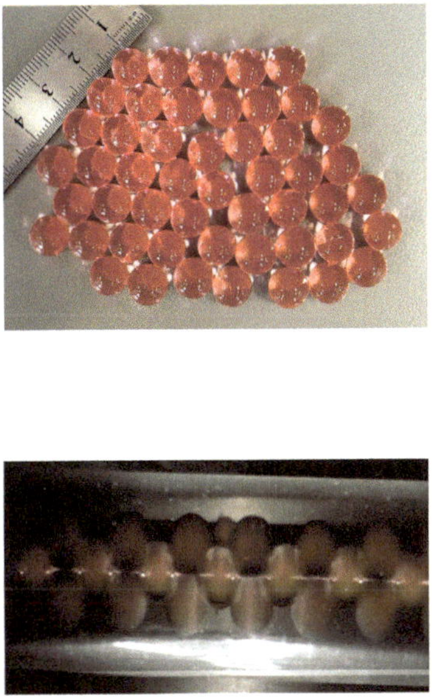

(b)

Figure 8.1 Experiments on rotational columnar structures of soft spheres can be performed using hydrogel spheres. Figure (a) shows a sample of such hydrogel spheres. Figure (b) shows a $(4, 2, 2)$ uniform structure of hydrogel spheres as obtained experimentally from the rapid rotations of a lathe.

hydrophilic polymer networks that, when in contact with water, swell up to diameters of $d_{\text{sphere}} \approx 1$ cm [see the scale in Figure 8.1(a)]. Such a system of soft spheres can be produced by immersing hydrophilic-polymer-based colloidal particles in water. The hydrogel spheres, after absorbing water, become elastic and deformable. The density of hydrogel spheres is slightly higher than that of water. This can be solved by adding salt to the water medium or choosing another type of liquid.

In fact, as shown in Figure 8.1(b), we have already performed preliminary experiments with hydrogel spheres. In these experiments, we placed the hydrogel spheres in a glycerol solution, which has a significantly higher density than the hydrogel spheres. Rapid rotations of the lathe lead to the formation of a variety of structures, such as the $(4, 2, 2)$ uniform structure in Figure 8.1(b).

However, there exist two major technical difficulties that remain to be overcome. As it can be seen from Figure 8.1(a), the hydrogel spheres are slightly polydisperse because not every sphere has absorbed the same amount of water during its formation. This problem might be overcome by filtering the hydrogel spheres by size. The second problem concerns a density increase of the spheres as a result of glycerol absorption. A possible way out is to make the surface of every sphere glycerol-proof.

8.5 Hard-Sphere Chains in a Cylindrical Harmonic Potential

8.5.1 Summary

The simplest possible columnar structure is a linear chain of spheres. Such a structure, confined by a radial harmonic potential and by hard walls at both ends, exhibits a variety of buckled structures as it is compressed longitudinally. As discussed in Chapter 7, such structures have been studied experimentally as well as theoretically.

Such buckled structures of hard spheres can be generated experimentally by Lee *et al.*'s method, where polymeric spheres are placed inside a rotating liquid-filled tube [Lee *et al.* (2017)]. For such experiments, we have carried out corresponding numerical simulations based on an iterative stepwise method, and we have developed a theoretical analysis based on a linear approximation. A variety of structures that emerge through bifurcations have been predicted and presented in energy bifurcation diagrams (Figure 7.6). The predicted stable structures have been observed in simulations and experiments.

8.5.2 Outlook: Extensions of Current Simulations and Experiments

Our simulations and experiments can be extended in a variety of possible directions. The iterative stepwise method, which is based on conditions of force balance, does not work well in the large-compression regimes. This problem may be overcome by adopting a more sophisticated root-searching method, for example,

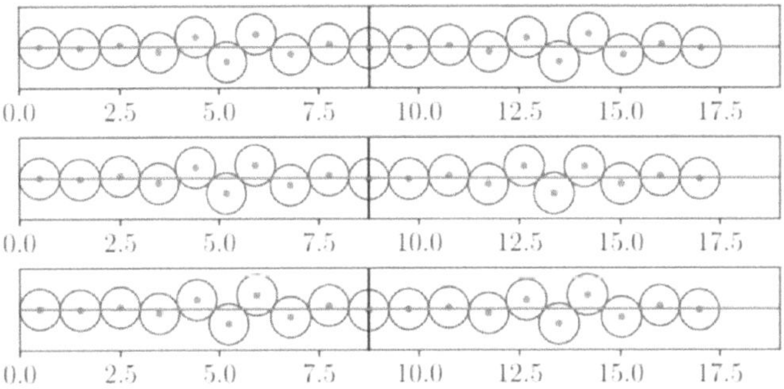

Figure 8.2 (Colour online) Three preliminary examples of double-peak high-energy structures. For each example, the black vertical line indicates the midpoint, and the blue horizontal line is the rotational axis. All structures were obtained from the iterative stepwise method.

the Newton–Raphson method. Our current approach, however, can readily be used to obtain high-energy structures that are not discussed in Chapter 7. They, however, appear in parts of Figure 7.5 where each line represents a particular structure, including such high-energy structures.

While, for each structure discussed in Chapter 7, there exists only a single peak in the corresponding displacement profile, the profile of some high-energy structures may consist of multiple peaks. Some preliminary examples as generated from the iterative stepwise method are shown in Figure 8.2, where the rotational energy of a double-peak structure is approximately twice that of its single-peak counterpart.

Simulations and experiments of such structures can also be extended to soft spheres, where the iterative stepwise method can be modified to allow inter-sphere overlaps. The results can then be compared directly to those of soft-sphere simulations that are based on the principle of energy minimisation (Subsection 7.2.2). Lathe experiments for soft spheres can be carried out using bubbles or hydrogel spheres (Figure 8.1). The problems of polydispersity and density as discussed in Subsection 8.4.2 would also be present for such soft-sphere experiments. In addition, spheres made of photoelastic materials [Majmudar and Behringer (2005); Daniels

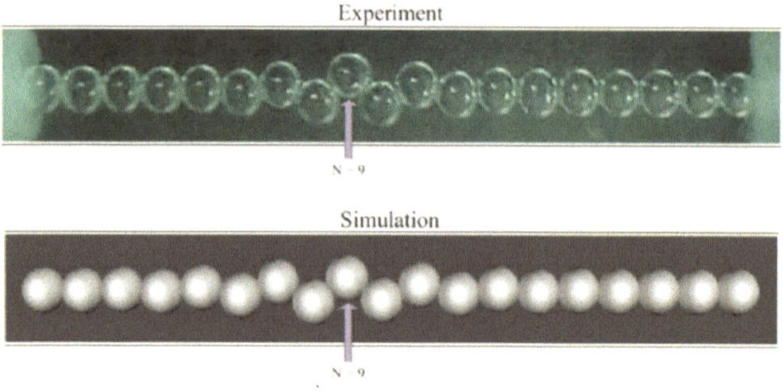

Figure 8.3 Top image: A modulated zigzag structure of a chain of 19 soap bubbles compressed between two stoppers at the ends of a liquid-filled tube, where the bubbles are pushed by buoyancy against the curved wall of the tube. Bottom image: The same modulated zigzag structure as obtained from soft-sphere simulations that are based on energy minimisation (Subsection 7.2.2). The chain of soft spheres was confined inside a cylindrical harmonic potential, and was compressed between two walls that are separated by a distance L apart. Figure reprinted, with permission, from Ref. [Weaire *et al.* (2020)].

et al. (2017)] can be used, where the magnitudes of inter-particle compressive forces (which play a vital role in the iterative stepwise method) can be visualised.

The work presented in Chapter 7 has also already in parts been extended to an experimental method using monodisperse bubbles confined in a horizontal liquid-filled tube. For a chain of monodisperse bubbles confined in a horizontal liquid-filled tube (Figure 8.3), buoyancy pushes the bubbles towards the curved tube wall. There they are confined by the walls at the ends of the tube. Being compressed along the axial direction, the chain of bubbles starts to buckle and a modulated zigzag structure similar to those described in Section 7.3 is formed.

A preliminary example of such a modulated zigzag structure of bubbles is presented in the top image of Figure 8.3. Such a structure can directly be compared to the structure obtained from soft-sphere simulations based on energy minimisation (Subsection 7.2.2), as shown in the bottom image of Figure 8.3. These experiments are mainly for illustration purposes. In general, the softness of particles may lead to some novel structures [Weaire *et al.* (2020)] that are not expected for hard-sphere packings [Hutzler *et al.* (2020)].

8.6 Soft-Disk Packings Inside a Two-Dimensional Rectangular Channel

Although the problem of packing spheres inside a cylinder can be understood as a problem of packing disks on the unrolled surface of the cylinder, the columnar structures addressed in this monograph are still three-dimensional in nature. The two-dimensional counterpart of this problem is the packing of circular disks inside a rectangular channel. Such two-dimensional packings have been studied in details for hard circular disks [Füredi (1991); Lubachevsky and Graham (2003)].

In our case, preliminary investigations have been carried out for soft-disk packings. The inter-particle interactions are modelled in the same way as those for soft spheres (as explained in Chapter 3). Similarities exist between the densest columnar structures of soft disks in confinement by a rectangular channel and the columnar structures of spheres discussed in this monograph. In both cases, we either have a hexagonal structure [Figure 8.4(a)], which is the two-dimensional counterpart of a uniform structure (Subsection 3.3.1), or a structure with a loss of contacts [Figure 8.4(b)], which is the two-dimensional counterpart of a line-slip arrangement (Subsection 3.3.2).

There are many quasi-two-dimensional systems where such packings are of great interest. One prominent example concerns the field of microfluidics. This is an emerging multi-disciplinary field that is concerned with the behaviour of fluids inside channels with a size of the microscale [Drenckhan *et al.* (2005)]. It finds applications in diagnostic medicine and microelectronics, where the precise manipulation of small amounts of fluid is important. One goal of researchers in this field is to control the flow of droplets or bubbles in microfluidic systems by varying the channel size or the system's pressure, where the corresponding structures of droplets or bubbles play a vital role in such investigations.

A two-dimensional bubble raft bounded by two plates, such as the one of monodisperse bubbles displayed in Figure 8.4(c),

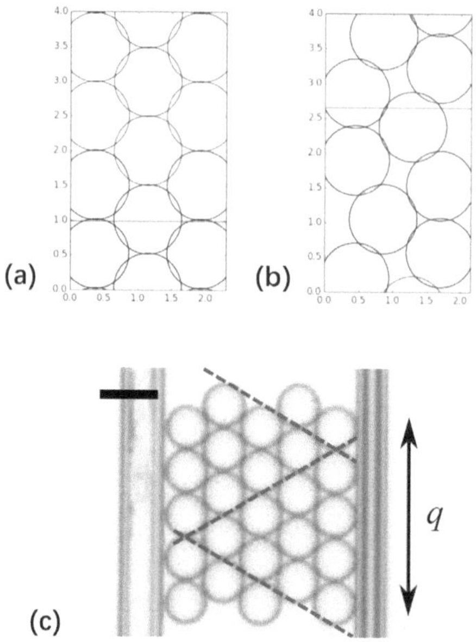

Figure 8.4 Soft-disk packings in simulations and experiments. For simulated packings of soft disks confined within a rectangular channel, Figure (a) shows an example of hexagonal packings, and Figure (b) shows an example of a line-slip-like structure. Figure (c) displays the optical microscopy image of a hexagonal crystal of buoyant droplets bounded by two glass plates [Ono-dit-Biot *et al.* (2020)].

can serve as a model system for crystal-to-glass transitions [Ono-dit-Biot *et al.* (2020)]. In a study by Ono-dit-Biot *et al.*, two-dimensional finite aggregates of oil droplets were compressed between two plates and their rearrangement under compression examined. By studying aggregates of mono- and polydisperse droplets, it gave insight into the energy landscape from crystals to glasses.

8.7 Limitations of the Soft-Sphere Model

In this section, we elaborate on the limitations of the soft-sphere model, which is the basis of all the soft-sphere simulations described in this monograph. By questioning the accuracy of the soft-sphere model and exploring other methods of modelling deformable objects, novel features of columnar structures may be discovered. We use the soft-disk model, which is a two-dimensional version of the soft-sphere model, to illustrate the corresponding theoretical limitations. Apart from the lack of volume conservation, there exist significant discrepancies in the average contact number Z between the soft-disk model and a simulated two-dimensional foam [Winkelmann *et al.* (2017)].

8.7.1 Soft Disks vs. Two-Dimensional Foams

To investigate the limitations of the soft-disk model, we compare results from this model with those from a well established model of two-dimensional foams, where the latter can be simulated using the `Plat` software [Bolton and Weaire (1991, 1992); Bolton (1996)]. The foam simulations described in this section were carried out by Friedrich Dunne, a former PhD Student in the Foams and Complex System Group of Trinity College Dublin. The software `Plat` can be used for the simulations of random two-dimensional foams [Bolton and Weaire (1991, 1992); Bolton (1996)]. Such simulations are not based on an energy-minimisation routine but are based on Plateau's laws. The films and liquid–gas interfaces of a two-dimensional foam are modelled as circular arcs, which are constrained to meet smoothly at vertices. The radius of curvature, r, of each arc is determined by the Young–Laplace equation. For a film with a surface tension γ, the Young–Laplace equation takes the form of $p_i - p_j = 2\gamma/r$, where p_i and p_j are the pressures in the adjacent bubbles i and j. For a liquid–gas interface, this law is given by $p_i - p_b = \gamma/r$, where p_b is the pressure in the Plateau border, set to be equal for all Plateau borders.

Two-dimensional (nearly)-dry-foam samples were generated by standard procedures [Bolton and Weaire (1991, 1992); Dunne *et al.* (2017)]: A random Delauney tessellation was used to compute a Voronoi network. It is then converted to an (as yet unequilibrated) dry foam, by a decoration of its vertices with small three-sided Plateau borders. Under the constraints that the arcs must meet smoothly and that the area of each bubble is conserved, the

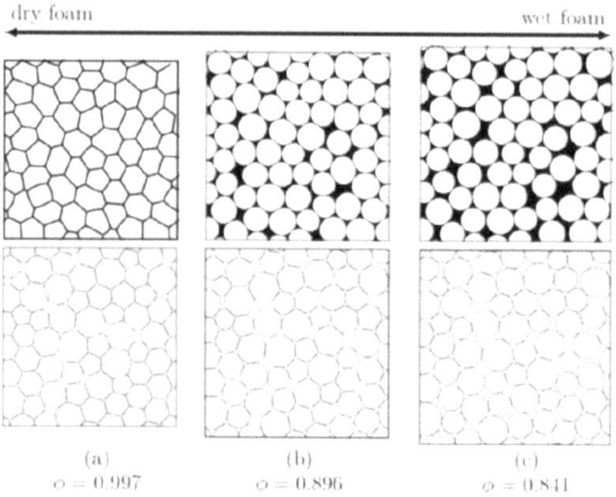

Figure 8.5 Comparison between two-dimensional foams (top images) and soft-disk packings (bottom images) at packing fractions of (a) $\phi = 0.997$, (b) $\phi = 0.896$ and (c) $\phi = 0.841$.

cell pressure and the vertex positions (x_n, y_n) are adjusted during the equilibration of the decorated Voronoi network. Equilibrium is reached when the change in the vertex positions has become negligibly small.

A progressive decrease of the packing fraction ϕ, in steps of $\Delta\phi = 0.001$, to a particular value of ϕ was implemented, with the system equilibrated at each step. The packing fraction was decreased by a proportional reduction of the bubble areas. The bubble-radius distribution, which is calculated from the bubble areas, follows a log-normal distribution. For a comparison at a similar polydispersity, random soft-disk packings were generated via the BFGS minimisation routine (Subsection 3.5.1) [Byrd *et al.* (1995)].

Three samples of $N = 60$ bubbles at different packing fractions are presented in the top images of Figure 8.5. The bottom images show the corresponding soft-disk packings at the same values of ϕ. In the soft-disk packings, the disks simply overlap for increasing packing fraction. But in the simulations of two-dimensional foams, the bubbles undergo a deformation into polygons as the dry-foam limit of $\phi = 1$ is approached.

8.7.2 Average Contact Number $Z(\phi)$

A quantitative discrepancy in the contact number $Z(\phi)$ exists between soft-disk packings and two-dimensional foams. Previous simulations [Durian (1995); O'Hern *et al.* (2002, 2003)] showed that the average contact number $Z(\phi)$ of the soft-disk model varies as

$$Z(\phi) - Z_c \propto (\phi - \phi_c)^{1/2}, \tag{8.1}$$

where ϕ_c is the packing fraction at the jamming point and Z_c is the critical contact number at ϕ_c. While the condition of local stability requires the presence of at least *three* neighbours for each disk at ϕ_c, the condition of global stability requires the critical contact number to be $Z_c = 4$ for an infinitely large two-dimensional system [van Hecke (2010)]. As a finite-size correction for our finite-sized systems with periodic boundaries, the critical contact number is given by

$$Z_c = 4\left[1 - \frac{1}{N}\right], \tag{8.2}$$

where N is the number of bubbles. A system with $N = 60$ bubbles corresponds to a critical contact number of $Z_c = 3.933$. This relation is obtained as follows: For a periodic system of N bubbles, we can fix the position of one bubble without any loss of generality, leaving only $(N - 1)$ bubbles free to move. The number of degrees of freedom, $2(N - 1)$, is then equated with the number of contacts, $ZN/2$. As a standard practice [O'Hern *et al.* (2003); Katgert and van Hecke (2010)], rattlers, which are bubbles with less than three contacts, were excluded in our theoretical analysis. Such bubbles, which are mechanically unstable, are not part of the connected network.

The main plot in Figure 8.6 displays the average contact number $Z(\phi)$ as a function of the packing fraction ϕ for two-dimensional foams (red dots) and soft-disk packings (bluish green dots), for a system size of $N = 60$ bubbles or disks. The soft-disk data displays an expected square-root behaviour, as indicated by the fit with the bluish solid curve. But for two-dimensional foams, the simulation data fit well to a linear relation between $Z(\phi)$ and ϕ at $\phi > \phi_c$, as

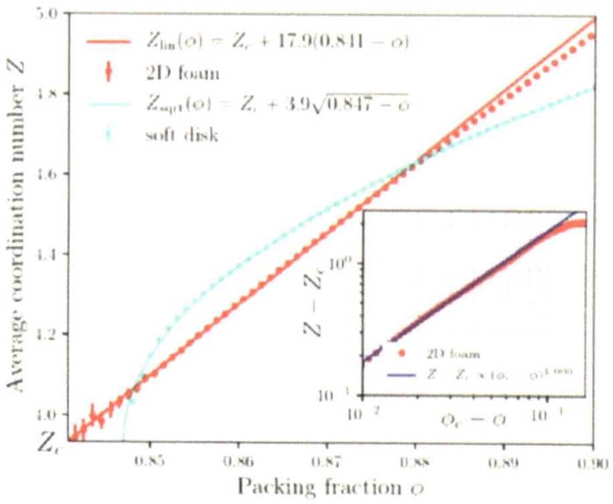

Figure 8.6 Comparison of the average contact number Z between two-dimensional foams (red dots) and soft-disk packings (bluish green dots), for a system size of $N = 60$ bubbles or disks. For two-dimensional foams, the average contact number Z varies linearly with $(\phi - \phi_c)$ in the vicinity of $\phi_c = 0.841 \pm 0.001$, as obtained from an averaging of over 600000 realisations. This linear relation is confirmed by the corresponding log–log plot in the inset. For soft-disk packings, the value of $Z(\phi)$ exhibits a square-root behaviour, as obtained from an averaging of over 20000 realisations.

indicated by the red solid curve in Figure 8.6). Fitting

$$Z(\phi) = Z_c + k_f(\phi - \phi_c), \tag{8.3}$$

to the `plat` data of two-dimensional foams with $Z_c = 4 - (1/15)$ yields $k_f = 17.9 \pm 0.1$ and $\phi_c = 0.841 \pm 0.001$. This value of ϕ_c is in agreement with existing numerical results [Bideau and Troadec (1984); Katgert and van Hecke (2010); Dunne *et al.* (2017)]. In the inset of Figure 8.6 is a log–log plot of $[Z(\phi) - Z_c]$ against $(\phi - \phi_c)$. The value of ϕ_c was adjusted to yield the best *linear* relationship in the log–log plot. This corresponds to a power-law relation of

$$Z(\phi) - Z_c \propto (\phi - \phi_c)^{\alpha}, \tag{8.4}$$

where $\phi_c = 0.841 \pm 0.001$ and $\alpha = 1.000 \pm 0.004$, hence a linear relation between $Z(\phi)$ and ϕ.

If one describes foams in the wet limit as packings of disks (or spheres), it is tempting to extend this analogy also to the functional relationship for $Z(\phi)$ and thus expect the same square-root relationship, at least to a first-order approximation. However, Surface Evolver simulations have shown that, for a two-dimensional foam, the bubble–bubble interactions are harmonic but not pairwise-additive [Lacasse *et al.* (1996)]. That is, the inter-particle interactions in the soft-disk model do not represent realistic bubble–bubble interactions. For three-dimensional foams, our results also suggest a deviation from the square-root behaviour of $Z(\phi)$. Apart from being not pairwise-additive, the bubble–bubble interactions of a three-dimensional foam are also not harmonic. Such interactions scale with the force f between droplets as $f^2 \ln(1/f)$, as first predicted by Morse and Witten [Morse and Witten (1993); Lacasse *et al.* (1996)].

The dependence of $Z(\phi)$ on ϕ is only one aspect in which there exist discrepancies between soft-particle packings and foam systems. Other discrepancies between the two types of systems have also been studied [Winkelmann *et al.* (2017)]. Although the physical implications of such discrepancies for columnar structures remain unclear, it suggests the need of a more suitable theoretical model for investigations of equilibrium structures of foams or emulsions. One such model is discussed in the next section.

8.8 The Morse–Witten Model for Deformable Spheres

Based on the Morse–Witten theory [Morse and Witten (1993)], the Morse–Witten model [Morse and Witten (1993); Höhler and Cohen-Addad (2017); Weaire *et al.* (2017); Dunne *et al.* (2019)] simulates deformable spheres such as bubbles or droplets. This theory describes the shape of a bubble subject to a point force and an equal but opposite body force, where the latter can be buoyancy or gravity. Figure 8.7 displays the shape of such a deformed bubble for the cases of (a) a three-dimensional bubble and (b) a two-dimensional bubble, as predicted by the Morse–Witten theory.

A review on simulations based on the Morse–Witten theory in provided by Höhler and Cohen-Addad [Höhler and Cohen-Addad (2017)]. As illustrated in Figure 8.7(a), this theory can be used to describe the equilibrium shape of a single bubble pressed against a horizontal plate by an external body force. The contact force between

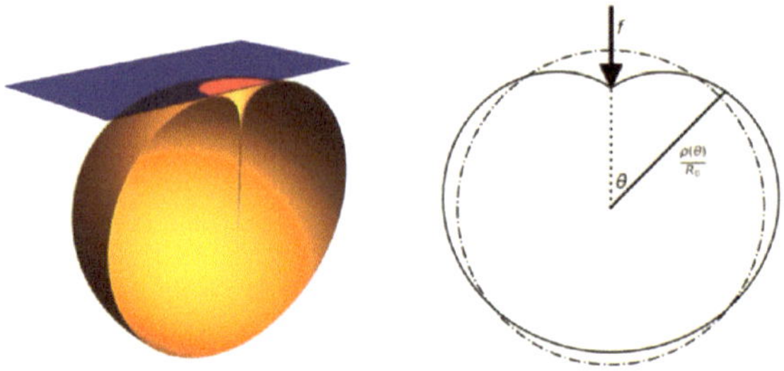

(a) (b)

Figure 8.7 The Morse–Witten theory is used to estimate the shape of a deformed bubble. In Figure (a), the shape of a three-dimensional bubble pushed by buoyancy against a horizontal plate (blue) is displayed. The red disk at the top indicates the contact area. The yellow part below the red disk is disregarded because the corresponding volume is negligible compared to the rest of the bubble [Höhler and Cohen-Addad (2017)]. Figure (b) shows the shape of a two-dimensional bubble deformed by a point force f and an equal but opposite body force, as predicted by the Morse–Witten theory. For the purpose of comparison, the undeformed bubble, which takes the shape of a disk of radius R_0, is indicated by the dashed circle. The sliver right below the faint horizontal dashed line is assumed to be negligible [Dunne *et al.* (2019)]. This justifies the use of a point force to model the bubble contact.

the plate and the bubble is assumed to be a point force f. The Morse–Witten theory approximates the bubble shape by linearizing the interfacial curvature C in the Young–Laplace equation

$$\gamma C = \Delta p\,. \tag{8.5}$$

where γ is the interfacial surface tension and Δp is the pressure difference across the interface. For a three-dimensional bubble, this approximation leads to a shape divergence at the point of contact, as illustrated in Figure 8.7(a) by the yellow part below the red disk, where the correcting volume is negligible. The shape of a two-dimensional bubble can be estimated in a similar fashion [Figure 8.7(b)], with the difference being that no divergence occurs at $\theta = 0$. The comparison between the dashed circle and the shape of the bubble in Figure 8.7(b) is a clear illustration of the effects of deformation.

While the soft-sphere model does not consider such deformation and assumes interactions to be pairwise-additive, the Morse–Witten theory applies to non-pairwise-additive bubble–bubble interactions where the corresponding effects of deformation are taken into account. A first simulation of a polydisperse wet foam was implemented by Dunne *et al.* [Dunne *et al.* (2019)]. This two-dimensional simulation, which considered all contact forces between neighbouring bubbles, was found to be accurate in the wet limit where the bubbles are all spherical in shape. However, simulations based on the Morse–Witten model have only been carried out for two-dimensional foams. In the context of columnar structures, a three-dimensional implementation of the Morse–Witten theory needs to be developed.

A two-dimensional implementation of the Morse–Witten model can be used to extend our work on the planar structures of sphere chains in a cylindrical harmonic potential (Chapter 7) to systems of monodisperse bubbles. The centripetal force on each bubble needs to be added, and at both ends of the rotating tube the bubble-wall interactions can be modelled in terms of horizontally aligned point forces.

It can also be used to study other two-dimensional packing problems, such as those introduced in Section 8.6. The Morse–Witten theory can be used to study bubbles compressed between two plates (considering the corresponding bubble-wall interactions). The resulting structures should give a close comparison to structures from the application of droplets or bubbles in microfluidic systems.

A three-dimensional implementation of the Morse–Witten model can be used to extend the research described in Chapters 5 and 6. However, huge challenges would have to be overcome for using this model to simulate columnar foams in cylindrical confinement. One major challenge would be to model accurately the contact geometry between a bubble and the curved cylindrical wall.

Appendix A

Tabulated Hard Sphere Results

In this appendix, we recapitulate some of the results of hard-sphere packings as reported by Mughal *et al.* [Mughal *et al.* (2012)].

Table A.1 provides a list of densest packings of hard spheres for $D \in [1, 2.71486]$, where $D \equiv D_{\text{cylinder}}/d_{\text{sphere}}$ is the cylinder-to-sphere diameter ratio. For each structure, the corresponding range of D and the number of spheres in a unit cell are given. The spheres of any structure listed in the table are all in contact with the cylindrical wall. The table does not take into account cases of $D > 2.71486$ where the densest structures consist of internal spheres. It includes all the structures that are indicated in the phase diagram of Figure 5.2.

For disk packings with a periodicity vector $\vec{V}$, Figure A.1 shows a plot of the surface density as a function of $|\vec{V}|$. Figure A.2 shows a comparison between accurate packing-fraction data for the packings of hard spheres in cylindrical confinement (upper black solid curve) and approximate packing-fraction data as computed from the affinely stretched disks on the unrolled surface of a cylinder (lower black dashed curve). Both figures are referred to frequently in Chapter 5 for a discussion of structural transitions. Such information has been used in the construction of the directed network in Figure 5.10 of Subsection 5.4.3.

Table A.1 List of densest packings of hard spheres for $D \in [1, 2.71486]$. At $D > 2.71486$, densest structures with internal spheres emerge. Bold numerals indicate the corresponding types of line-slip structures. The table is taken from Ref. [Mughal *et al.* (2012)].

structure	range	spheres in unit cell	notation
1 (C_1)	$D = 1$	1	straight chain
2	$1 \leq D \leq 1.866$	1	zigzag
3	$1.866 \leq D \leq 1.995$	1	twisted zigzag
4	$1.995 \leq D < 2.0$	2	(**2**,1,1)
5 (C_2)	$D = 2.0$	2	(2,2,0)
6	$2.0 < D \leq 2.039$	2	(2,**2**,0)
7	$D = 2.039$	1	(3,2,1)
8	$2.039 \leq D \leq 2.1413$	2	(3,**2**,1)
9	$2.1413 \leq D < 2.1545$	3	(**3**,2,1)
10 (C_3)	$D = 2.1547$	3	(3,3,0)
11	$2.1547 < D \leq 2.1949$	3	(3,**3**,0)
12	$2.1949 \leq D \leq 2.2247$	2	(3,**2**,1)
13	$D = 2.2247$	2	(4,2,2)
14	$2.2247 \leq D \leq 2.2655$	2	(4,2,**2**)\(4,**2**,2)
15	$2.2655 \leq D \leq 2.2905$	3	(3,**3**,0)
16	$D = 2.2905$	1	(4,3,1)
17	$2.2905 \leq D \leq 2.3804$	3	(4,**3**,1)
18	$2.3804 \leq D < 2.413$	4	(**4**,3,1)
19 (C_4)	$D = 2.4142$	4	(4,4,0)
20	$2.4142 < D \leq 2.4626$	4	(4,**4**,0)
21	$2.4626 \leq D \leq 2.4863$	3	(4,**3**,1)
22	$D = 2.4863$	1	(5,3,2)
23	$2.4863 \leq D \leq 2.5443$	3	(5,**3**,2)
24	$2.5443 \leq D \leq 2.5712$	4	(4,**4**,0)
25	$D = 2.5712$	1	(5,4,1)
26	$2.5712 \leq D \leq 2.655$	4	(5,**4**,1)
27	$2.655 \leq D < 2.7013$	5	(**5**,4,1)
28 (C_5)	$D = 2.7013$	5	(5,5,0)
29	$2.7013 < D \leq 2.71486$	5	(5,**5**,0)

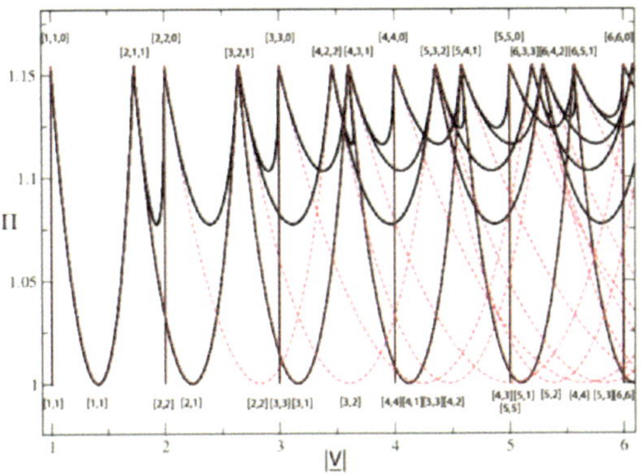

Figure A.1 Surface density as a function of $|\vec{V}|$, for disk packings with a periodicity vector $\vec{V}$. Each density peak corresponds to a uniform structure labelled with a particular phyllotactic notation (l, m, n). Each black solid curve connecting two peaks corresponds to a particular line-slip structure. The red dashed curves correspond to affine transitions that are outside the scope of this monograph. The plot was produced by Mughal *et al.* [Mughal *et al.* (2012)].

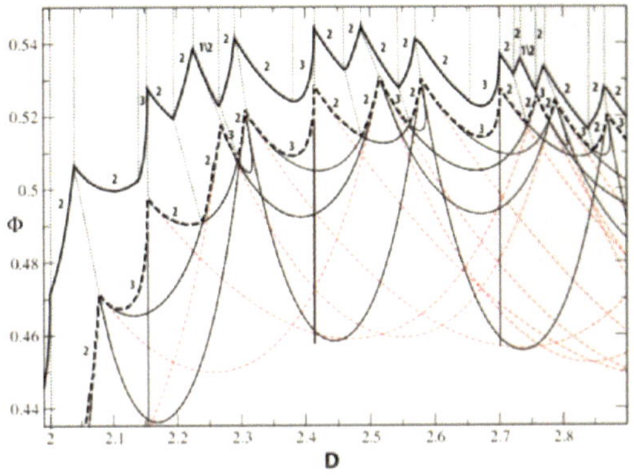

Figure A.2 Comparison between accurate packing-fraction data for the packings of hard spheres in cylindrical confinement (upper black solid curve) and approximate packing-fraction data as computed from the affinely stretched disks on the unrolled surface of a cylinder (lower black dashed curve). The dotted lines indicate the structural correspondence between the two curves. The plot was produced by Mughal *et al.* [Mughal *et al.* (2012)].

Appendix B

Minimisation Routines

For the minimisation of energy or enthalpy, the BFGS method is employed for any search of local minima, and the Basin-hopping method is employed for any search of global minima. Both methods are part of the scipy library [Jones *et al.* (2001)] python programming language.

In physics, stable structures are obtained by searching for the minimum of a thermodynamic potential. The energy and the enthalpy are two examples of such a potential. In this appendix, the function to be minimised is generally referred to as the energy function $\mathrm{E}(\vec{X})$. The N parameters, on which the energy function depends, are all contained within the vector $\vec{X}$. These parameters are mainly the positional coordinates of spheres in cylindrical confinement.

The BFGS Method: A Quasi-Newton Method

The simplest algorithm that can be employed to find the local minimum of a function is the gradient-descent (or steepest descent) algorithm. The local minimum is attained through iterative steps along the directions of the force vector

$$\vec{F} = -\vec{\nabla}\mathrm{E}(\vec{X}) = -\left(\frac{\partial \mathrm{E}}{\partial X_1}, \frac{\partial \mathrm{E}}{\partial X_2}, \ldots\ldots \frac{\partial \mathrm{E}}{\partial X_N}\right), \tag{B.1}$$

where N is the length of the parameter vector $\vec{X}$.

We used a more sophisticated algorithm to search for the local minimum: The BFGS method [Byrd *et al.* (1995); Nocedal and Wright (1999); Jones *et al.* (2001)]. This method, which is named after its discoverers Broyden, Fletcher, Goldfarb, Shanno, belongs to the class of quasi-Newton algorithms [Nocedal and Wright (1999)]. Quasi-Newton algorithms are advancements of the conjugate-gradient algorithm. Such algorithms are sometimes also referred to as conjugate gradients, even though there are technical differences between quasi-Newton algorithms and conjugate gradients [Nocedal and Wright (1999)].

Any Newton method is based on the second-order Taylor expansion around the minimum $\vec{X}_0$ of an energy function

$$\mathrm{E}(\vec{X}) = \mathrm{E}(\vec{X}_0) - \vec{F}^{\mathrm{T}} \cdot (\vec{X} - \vec{X}_0) + \frac{1}{2}(\vec{X} - \vec{X}_0)^{\mathrm{T}} \cdot \vec{H} \cdot (\vec{X} - \vec{X}_0), \quad \text{(B.2)}$$

where $\vec{F}$ is the force vector and $\vec{H}$ is the Hessian of $\mathrm{E}(\vec{X})$. The Hessian is an $N \times N$ matrix that contains all second derivatives of the energy:

$$\vec{H} = \begin{pmatrix} \frac{\partial^2 \mathrm{E}}{\partial X_1 \partial X_1} & \frac{\partial^2 \mathrm{E}}{\partial X_1 \partial X_2} & \cdots & \frac{\partial^2 \mathrm{E}}{\partial X_1 \partial X_N} \\ \frac{\partial^2 \mathrm{E}}{\partial X_2 \partial X_1} & \frac{\partial^2 \mathrm{E}}{\partial X_2 \partial X_2} & \cdots & \vdots \\ \vdots & \vdots & \ddots & \vdots \\ \frac{\partial^2 \mathrm{E}}{\partial X_N \partial X_1} & \cdots & \cdots & \frac{\partial^2 \mathrm{E}}{\partial X_N \partial X_N} \end{pmatrix}. \quad \text{(B.3)}$$

To avoid complicated analytic calculations, this matrix is often numerically approximated. Any Newton method that numerically approximates the Hessian is called a quasi-Newton algorithm.

Figure B.1 displays a pseudo-code for quasi-Newton methods. The main step (line 4 of Figure B.1) of the iterative scheme can be derived from Equation (B.2). By setting $E(\vec{X}_0) = 0$ (WLOG), the equation

$$\vec{H} \cdot \underbrace{(\vec{X} - \vec{X}_0)}_{\vec{p}_0} = \vec{F} \quad \text{(B.4)}$$

is satisfied in the local minimum of the Taylor expansion. Thus, solving this equation iteratively for the direction $\vec{p}_0$ yields a downhill direction towards a possible minimum. This step is followed by a line

Algorithm 1: The quasi-Newton method

input : Energy function $\mathrm{E}(\boldsymbol{X})$ and initial condition $\boldsymbol{X}_0$ of length N
output: Local minimum $\boldsymbol{X}_{\min}$

$\boldsymbol{F}_0 = -\nabla \mathrm{E}(\boldsymbol{X}_0)$; // Calculate forces
$\boldsymbol{H}_0 = \boldsymbol{H}(\boldsymbol{X}_0)$; // Calculate Hessian
while $|\boldsymbol{F}_0| > \epsilon$ **do**
  Solve matrix equation $\boldsymbol{H}_0 \boldsymbol{p}_0 = \boldsymbol{F}_0$ for $\boldsymbol{p}_0$;
  Line search along $\boldsymbol{X}_0 + \alpha \boldsymbol{p}_0$ for α ;
  $\boldsymbol{X}_{\min} = \boldsymbol{X}_0 = \boldsymbol{X}_0 + \alpha \boldsymbol{p}_0$;
  $\boldsymbol{F}_0 = -\nabla \mathrm{E}(\boldsymbol{X}_0)$; // Update forces
  $\boldsymbol{H}_0 = \boldsymbol{H}(\boldsymbol{X}_0)$; // Update Hessian
end

Figure B.1 Pseudo-code for quasi-Newton methods, where the nearest local minimum of an energy function $\mathrm{E}(\vec{X})$ is searched for. The direction of $\vec{p}_0$ is found by solving the matrix equation (B.4) iteratively. This is followed by a line search to find the step α to the nearest minimum in this direction. The parameter vector $\vec{X}_{\min}$, the initial condition $\vec{X}_0$, as well as the forces (or gradient) and the Hessian are then updated accordingly. For quasi-Newton methods, the Hessian is approximated numerically. The iterative scheme terminates when the absolute value of the forces drops below a certain threshold ϵ.

search in the direction of $\vec{p}_0$, which is usually done with the Wolfe algorithm [Nocedal and Wright (1999)]. This line search yields the step size α required to find the nearest minimum along the direction of $\vec{p}_0$. The parameter vectors $\vec{X}_{\min}$ and $\vec{X}_0$ are then updated to this minimum, and the forces and Hessian are calculated for this new minimum. The algorithm terminates when the absolute value of the forces $|\vec{F}_0|$ are below a certain threshold value ϵ, which is close to zero.

A common problem of this iterative scheme is the occurrence of a singular Hessian. If the Hessian is singular, the matrix equation (B.4) has no solution. The BFGS method circumvents this problem by directly approximating the inverse of the Hessian.

Any Newton method or conjugate-gradient-type algorithm is a deterministic algorithm, i.e., the algorithm, which does not contain any stochastic element, will always arrive at the same minimum for any given initial condition $\vec{X}_0$. It is therefore a reliable algorithm for any search of nearest local minima. We make use of this property to investigate structural transitions (Chapter 5). On the downside,

Figure B.2 (Colour online) Schematic illustration of a rugged energy landscape with many local minima [Wales and Doye (1997)]. The dashed lines map out the basins around each minimum. The red dot represents the global minimum, while the blue dot represents the local minimum that is nearest to the initial guess (black dot).

these algorithms are not suitable for searching for the global energy minimum because they easily get stuck in a local minimum.

The Basin-hopping algorithm

To find the global minimum of an energy function $E(\vec{X})$, a stochastic component is needed. The landscape of a one-dimensional energy function with many local minima is illustrated in Figure B.2. Any search via the use of a deterministic algorithm, such as the previously described BFGS method, would end up being stuck in the nearest local minimum (blue dot). A stochastic component makes it possible for the algorithm to jump out of a basin and reach the global minimum (red dot). The Basin-hopping algorithm is one of such stochastic global-minimisation routines [Wales and Doye (1997)]. Its iterative scheme resembles that of simulated annealing [Jones *et al.* (2001)]. Similar to the case of simulated annealing, the Basin-hopping algorithm is suitable for rugged energy landscapes with many local minima, such as the one in Figure B.2. It combines heuristic procedures with local searches, so as to effectively explore the entire parameter space.

Algorithm 1: Basin-hopping algorithm

input : Energy function $\mathrm{E}(\boldsymbol{X})$ with N-dimensional parameter $\boldsymbol{X}$
output: Global minimum $\boldsymbol{X}_{\min}$

$\boldsymbol{X}_0$ = random point in the parameter space ;
$\boldsymbol{X}_{\min} = \boldsymbol{X}_{\text{old}}$ = local minimum of $\mathrm{E}(\boldsymbol{X})$ around $\boldsymbol{X}_0$;
for $i = 0$ **to** niter **do**
  $\boldsymbol{X}'_0$ = randomly perturb parameter $\boldsymbol{X}_0$;
  $\boldsymbol{X}_{\text{new}}$ = local minimum of $\mathrm{E}(\boldsymbol{X})$ around $\boldsymbol{X}'_0$;
  q = random number $\in (0, 1]$;
  if $\mathrm{E}(\boldsymbol{X}_{\text{new}}) < \mathrm{E}(\boldsymbol{X}_{\text{old}})$ **or** $\exp(-(\mathrm{E}(\boldsymbol{X}_{\text{new}}) - \mathrm{E}(\boldsymbol{X}_{\text{old}}))/\mathrm{T}) < q$ **then**
    $\boldsymbol{X}_0, \boldsymbol{X}_{\text{old}} = \boldsymbol{X}'_0, \boldsymbol{X}_{\text{new}}$;   `// accept new step`
  end
  if $\mathrm{E}(\boldsymbol{X}_{\text{new}}) \leq \mathrm{E}(\boldsymbol{X}_{\min})$ **then**
    $\boldsymbol{X}_{\min} = \boldsymbol{X}_{\text{new}}$;   `// update new global minimum`
  end
end

Figure B.3 Pseudo-code for the basin-hopping algorithm. An energy function $\mathrm{E}(\vec{X})$ is minimised in N dimensions. The algorithm finds the global minimum iteratively by performing three steps in each of the `niter` iterations: (a) The parameter $\vec{X}_0$ is perturbed randomly and stored in $\vec{X}'_0$. (2) The nearest local minimum from $\vec{X}'_0$ is obtained through a conjugate-gradient-type algorithm. This step of perturbation is accepted according to the Metropolis acceptance criterion for temperature T. (3) The minimum $\vec{X}_{\min}$ is then updated accordingly.

The pseudo-code in Figure B.3 illustrates the iterative search scheme of the basin-hopping algorithm. The search is performed first by a perturbation of the system, followed by a local optimisation in a given number of steps, `niter`. With the initial guesses $\vec{X}'_0$, the local minima are calculated using conjugate-gradient- or quasi-Netwon-type algorithms (such as the BFGS method).

If the energy of the new local minimum is below the previous minimum, the new step is always accepted. Otherwise the step is accepted according to the Metropolis acceptance criterion, as in the case of simulated annealing. The step is accepted with the probability

$$p_A = \exp\left(-\frac{\mathrm{E}(\vec{X}_{\text{new}}) - \mathrm{E}(\vec{X}_{\text{old}})}{\mathrm{T}}\right), \tag{B.5}$$

where $\mathrm{E}(\vec{X}_{\text{old}})$ and $\mathrm{E}(\vec{X}_{\text{new}})$ are here the energies before and after the new step. With the temperature T, the likelihood of jumping to another basin can be tuned. During the iteration, the global minimum is updated accordingly.

The perturbations in each step of the algorithm ensure that more than just one local minimum is explored. With a sufficient number of iterations, the probability of finding the global minimum approaches unity. Like the case of simulated annealing, the acceptance rate of the random perturbations depends on the temperature T.

Appendix C

Energy of (*l*, *l*, 0) Structures

For $(l, l, 0)$ structures, such as the structures of $(2, 2, 0)$, $(3, 3, 0)$ and $(4, 4, 0)$, analytic energy expressions are derived in this appendix. The calculation starts with Equation (6.8) (Chapter 6) for the total energy per sphere, E^S_{rot}:

$$\frac{E^S_{\text{rot}}}{M\omega^2 {d_{\text{sphere}}}^2} = \frac{1}{2}\frac{R^2}{{d_{\text{sphere}}}^2} + \frac{1}{2}\frac{k}{M\omega^2}\left\langle\left(\frac{\delta_{ij}}{d_{\text{sphere}}}\right)^2\right\rangle . \tag{C.1}$$

The first term corresponds to the rotational energy, and the second term, which is generally structure-dependent, corresponds to inter-sphere interactions. An $(l, l, 0)$ structure is a stacking of layers of spheres, with each layer consisting of l spheres. This implies that the average $\left\langle \delta^2_{ij}/{d_{\text{sphere}}}^2 \right\rangle$ can be expressed as a sum of two components:

$$\left\langle\left(\frac{\delta_{ij}}{d_{\text{sphere}}}\right)^2\right\rangle = \Delta_{\text{Layer}} + 2\Delta_{\text{Top}} , \tag{C.2}$$

where

$$\Delta_{\text{Layer}} = \frac{[d_{\text{sphere}} - r_0(R)]^2}{{d_{\text{sphere}}}^2} \tag{C.3}$$

and

$$\Delta_{\text{Top}} = \frac{[d_{\text{sphere}} - h(L, R)]^2}{{d_{\text{sphere}}}^2} . \tag{C.4}$$

The term Δ_{Layer} corresponds to the overlap of a sphere with another sphere from the same layer, and the term Δ_{Top} corresponds to the overlap of a sphere with two other spheres from the adjacent layers above and below, respectively. R is the radial position of each sphere, and L is the inter-layer separation, where the latter is related to the linear number density ρ as described in Chapter 6. The term

$$r_0(R) = 2R\sin\left(\frac{\pi}{l}\right) \tag{C.5}$$

is the centre-to-centre separation within a layer. The inter-layer separation $h(L,R)$ can be derived from a consideration of two spheres at (x_0, y_0, z_0) and (x_1, y_1, z_1), respectively:

$$\begin{aligned} x_0 &= 0 & x_1 &= R\cos\left(\frac{\pi}{l}\right) \\ y_0 &= R\sin\left(\frac{\pi}{2}\right) & y_1 &= R\cos\left(\frac{\pi}{2}+\frac{\pi}{l}\right) \\ z_0 &= 0 & z_1 &= L\,. \end{aligned} \tag{C.6}$$

which yields

$$h(R,L) = \sqrt{R^2\cos^2\left(\frac{\pi}{l}\right) + R^2\left(\sin\left(\frac{\pi}{2}+\frac{\pi}{l}\right)-1\right)^2 + L^2} \tag{C.7}$$

Substituting these expressions for $r_0(R)$ and $h(R,L)$ into the expression for the total energy yields:

$$\frac{E^S_{(l,l,0)}(R,L)}{M\omega^2 {d_{\text{sphere}}}^2} = \frac{1}{2}\frac{R^2}{{d_{\text{sphere}}}^2} + \frac{1}{2}\frac{k}{M\omega^2 {d_{\text{sphere}}}^2}\left(\left(d_{\text{sphere}} - 2R\sin\left(\frac{\pi}{l}\right)\right)^2 + \left(d_{\text{sphere}} - \sqrt{R^2\cos^2\left(\frac{\pi}{l}\right) + R^2\left(\sin\left(\frac{\pi}{2}+\frac{\pi}{l}\right)-1\right)^2 + L^2}\right)^2\right). \tag{C.8}$$

The total energy E^S_{rot} is now only a function of two variables, the radial distance R and length L. With L (and all the other parameters) kept constant, the total energy E^S_{rot} can be minimised with respect to R by analytic or numerical means. The length L can be related to the linear number density ρ as described in Chapter 6.

Bibliography

[Abbaschian and Reed-Hill (2008)] R. Abbaschian and R. E. Reed-Hill, *Physical Metallurgy Principles* (Cengage Learning, 2008). https://books.google.de/books?id=wh4v6UWjYdIC

[Amir and Nelson (2012)] A. Amir and D. R. Nelson, *Dislocation-mediated growth of bacterial cell walls*, Proc. Natl. Acad. Sci. U.S.A. 109, 9833–9838. (2012).

[Aste and Weaire (2008)] T. Aste and D. Weaire, *The Pursuit of Perfect Packing*, 2nd Ed. (Taylor and Francis, London, 2008).

[Bautista-Carbajal *et al.* (2013)] G. Bautista-Carbajal, A. Moncho-Jordá, and G. Odriozola, *Further details on the phase diagram of hard ellipsoids of revolution*, J. Chem. Phys. 138, 064501 (2013).

[Beatus *et al.* (2006)] T. Beatus, T. Tlusty, and R. Bar-Ziv, *Phonons in a one-dimensional microfluidic crystal*, Nat. Phys. 2, 743–748 (2006).

[Bernal (1960)] J. D. Bernal, *Geometry of the structure of monatomic liquids*, Nature 185, 68–70 (1960).

[Bideau and Troadec (1984)] D. Bideau and J. P. Troadec, *Compacity and mean coordination number of dense packings of hard discs*, J Phys. C Solid State Phys. 17, L731–L736 (1984).

[Boltenhagen and Pittet (1998)] P. Boltenhagen and N. Pittet, *Structural transitions in crystalline foams*, EPL 41, 571–576 (1998).

[Boltenhagen *et al.* (1998)] P. Boltenhagen, N. Pittet, and N. Rivier, *Giant deformations and topological hysteresis of an ordered foam*, EPL 43, 690–694 (1998).

[Bolton (1996)] F. Bolton, *Software PLAT: A computer code for simulating two-dimensional liquid foams*, GitHub repository, https://github.com/fbolton/plat

[Bolton and Weaire (1991)] F. Bolton and D. Weaire, *The effects of Plateau borders in the two-dimensional soap froth. I. Decoration lemma and diffusion theorem.*, Philos. Mag. B 63, 795–809 (1991).

[Bolton and Weaire (1992)] F. Bolton and D. Weaire, *The effects of Plateau borders in the two-dimensional soap froth. II. General simulation and analysis of rigidity loss transition.*, Philos. Mag. B 65, 473–487 (1992).

[Borodavka *et al.* (2012)] A. Borodavka, R. Tuma, and P. G. Stockley, *Evidence that viral RNAs have evolved for efficient, two-stage packaging*, Proc. Natl. Acad. Sci. U.S.A. 109, 15769–15774 (2012).

[Boromand *et al.* (2018)] A. Boromand, A. Signoriello, F. Ye, C. S. O'Hern, and M. D. Shattuck, *Jamming of deformable polygons*, Phys. Rev. Lett. 121, 248003 (2018).

[Boromand *et al.* (2019)] A. Boromand, A. Signoriello, J. Lowensohn, C. S. Orellana, E. R. Weeks, F. Ye, M. D. Shattuck, and C. S. O'Hern, *The role of deformability in determining the structural and mechanical properties of bubbles and emulsions*, Soft Matter 15, 5854–5865 (2019).

[Brakke (1992)] K. A. Brakke, *The Surface Evolver*, Exp. Math. 1, 141–165 (1992).

[Brinkley (1997)] W. Brinkley, *Microtubules: A brief historical perspective*, J. Struct. Biol. 118, 84–86 (1997).

[Brisson and Unwin (1984)] A. Brisson and P. N. Unwin, *Tubular crystals of acetylcholine receptor*, J. Cell Biol. 99, 1202–1211 (1984).

[Bryan (1974)] J. Bryan, *Microtubules*, BioScience 24, 701-711 (1974).

[Byrd *et al.* (1995)] R. H. Byrd, P. Lu, J. Nocedal, and C. Zhu, *A limited memory algorithm for bound constrained optimization*, SIAM J. Sci. Comput. 16, 1190–1208 (1995).

[Cai *et al.* (2007)] W. Cai, U. K. Chettiar, A. V. Kildishev, and V. M. Shalaev, *Optical cloaking with metamaterials*, Nat. Photonics 1, 224–227 (2007).

[Chan (2011)] H.-K. Chan, *Densest columnar structures of hard spheres from sequential deposition*, Phys. Rev. E 84, 050302(R) (2011).

[Chan (2012)] H.-K. Chan, *The scientific adventure of packing*, ENGineering INSpiration (2012), `https://www.engins.org/2012/11/05/the-scientific-adventure-of-packing/` (Visited on 2020-09-23)

[Chan (2013)] H.-K. Chan, *A hybrid helical structure of hard sphere packing from sequential deposition*, Philos. Mag. 93, 4057–4069 (2013).

[Chan *et al.* (2019)] H.-K. Chan, Y. Wang, and H. Han, *Densest helical structures of hard spheres in narrow confinement: An analytic derivation*, AIP Adv. 9, 125118 (2019).

[Chopra *et al.* (1995)] N. G. Chopra, R. J. Luyken, K. Cherrey, V. H. Crespi, M. L. Cohen, S. G. Louie, and A. Zettl, *Boron nitride nanotubes*, Science 269, 966–967 (1995).

[Cong *et al.* (2014)] V. T. Cong, E.-O. Ganbold, J. K. Saha, J. Jang, J. Min, J. Choo, S. Kim, N. W. Song, S. J. Son, S. B. Lee, and S.-W. Joo, *Gold nanoparticle silica nanopeapods*, J. Am. Chem. Soc. 136, 3833–3841 (2014).

[Conway and Sloane (1999)] J. Conway and N. J. A. Sloane, *Sphere packings, lattices and groups*, 3rd Ed. (Springer, New York, 1999).

[Cox *et al.* (In preparation)] K. Cox, J. Winkelmann, and S. Hutzler, *Chainlike arrangement of equal-volume soap bubbles* (In preparation).

[Daniels *et al.* (2017)] K. E. Daniels, J. E. Kollmer, and J. G. Puckett, *Photoelastic force measurements in granular materials*, Rev. Sci. Instrum. 88, 051808 (2017).

[Delaney *et al.* (2005)] G. Delaney, D. Weaire, S. Hutzler, and S. Murphy, *Random packing of elliptical disks*, Philos. Mag. Lett. 85, 89–96 (2005).

[DesignerData (2019)] *DesignerData*, `http://www.designerdata.nl/plastics/thermo+plastics/PP` (Visited on 2019-08-13)

[Dessup *et al.* (2015)] T. Dessup, T. Maimbourg, C. Coste, and M. Saint Jean, *Linear instability of a zigzag pattern*, Phys. Rev. E 91, 022908 (2015).

[Donev *et al.* (2004a)] A. Donev, F. H. Stillinger, P. M. Chaikin, and S. Torquato, *Unusually dense crystal packings of ellipsoids*, Phys. Rev. Lett. 92, 255506 (2004).

[Donev *et al.* (2004b)] A. Donev, I. Cisse, D. Sachs, E. A. Variano, F. H. Stillinger, R. Connelly, S. Torquato, and P. M. Chaikin, *Improving the density of jammed disordered packings using ellipsoids*, Science, 303, 990–993 (2004).

[Drenckhan *et al.* (2005)] W. Drenckhan, S. J. Cox, G. Delaney, H. Holste, D. Weaire, and N. Kern, *Rheology of ordered foams–on the way to discrete microfluidics*, Colloids Surf. A 263, 52–64 (2005).

[Douglass *et al.* (2017)] I. Douglass, H. Mayger, T. Hudson, and P. Harrowell, *The stabilization of tubular crystals in mixtures of spherical particles*, Soft Matter 13, 1344–1351 (2017).

[Dunne *et al.* (2017)] F. F. Dunne, F. Bolton, D. Weaire, and S. Hutzler, *Statistics and topological changes in 2D foam from the dry to the wet limit*, Philos. Mag. 97, 1768–1781 (2017).

[Dunne *et al.* (2019)] F. F. Dunne, J. Winkelmann, D. Weaire, and S. Hutzler, *Implementation of Morse-Witten theory for a polydisperse wet 2d foam simulation*, Philos. Mag. 99, 2303–2320 (2019).

[Duran-Olivencia and Gordillo (2009)] F. J. Duran-Olivencia and M. C. Gordillo, *Ordering of hard spheres inside hard cylindrical pores*, Phys. Rev. E 79, 061111 (2009).

[Durian (1995)] D. J. Durian, *Foam mechanics at the bubble scale*, Phys. Rev. Lett. 75, 4780–4783 (1995).

[Erickson (1973)] R. O. Erickson, *Tubular packing of spheres in biological fine structure*, Science 181, 705–716 (1973).

[Erickson (1983)] R. O. Erickson, *The geometry of phyllotaxis* in *The growth and functioning of leaves* (Proceedings of a symposium held prior to the thirteenth International Botanical Congress at the University of Sydney 18-20 August 1981, edited by J. E. Dale and F. L. Milthrope) (Cambridge University Press, Cambridge, 2011), pp. 53–88.

[Fan and Govorov (2011)] Z. Fan and A. O. Govorov, *Helical metal nanoparticle assemblies with defects: Plasmonic chirality and circular dichroism*, J. Phys. Chem. C 115, 13254–13261 (2011).

[Fan *et al.* (2010)] J. A. Fan, C. Wu, K. Bao, J. Bao, R. Bardhan, N. J. Halas, V. N. Manoharan, P. Nordlander, G. Shvets, and Federico Capasso, *Self-assembled plasmonic nanoparticle clusters*, Science 328, 1135–1138 (2010).

[Finney (2013)] J. L. Finney, *Bernal's road to random packing and the structure of liquids*, Philos. Mag. 93, 3940–3969 (2013).

[Fu (2017)] L. Fu, *Soft self-assembly and densest packings in colloidal models*, PhD thesis (2017).

[Fu *et al.* (2016)] L. Fu, W. Steinhardt, H. Zhao, J. E. S. Socolar, and P. Charbonneau, *Hard sphere packings within cylinders*, Soft Matter 12, 2505–2514 (2016).

[Fu *et al.* (2017)] L. Fu, C. Bian, C. W. Shields IV, D. F. Cruz, G. P. López, and P. Charbonneau, *Assembly of hard spheres in a cylinder: A computational and experimental study*, Soft Matter 13, 3296–3306 (2017).

[Füredi (1991)] Z. Füredi, *The densest packing of equal circles into a parallel strip*, Discrete Comput. Geom. 6, 95–106 (1991).

[Hales (2005)] T. C. Hales, *A proof of the Kepler conjecture*, Ann. Math. 162, 1065–1185 (2005).

[Hansen and McDonald (1990)] J.-P. Hansen and I. R. McDonald, *Theory of Simple Liquids* (Elsevier, 1990).

[Heitkam *et al.* (2012)] S. Heitkam, W. Drenckhan, and J. Fröhlich, *Packing spheres tightly: Influence of mechanical stability on close-packed sphere structures*, Phys. Rev. Lett., 108, 148302 (2012).

[Hill *et al.* (2004)] J. P. Hill, W. Jin, A. Kosaka, T. Fukushima, H. Ichihara, T. Shimomura, K. Ito, T. Hashizume, N. Ishii, and T. Aida, *Self-assembled hexa-peri-hexabenzocoronene graphitic nanotube*, Science 304, 1481–1483 (2004).

[Hodak and Girifalco (2003)] M. Hodak and L. A. Girifalco, *Ordered phases of fullerene molecules formed inside carbon nanotubes*, Phys. Rev. B 67, 075419 (2003).

[Höhler and Cohen-Addad (2017)] R. Höhler and S. Cohen-Addad, *Many-body interactions in soft jammed materials*, Soft Matter 13, 1371–1383 (2017).

[Hull (1976)] R. Hull, *The structure of tubular viruses*, Adv. Virus Res. 20, 1–32 (1976).

[Hunt *et al.* (2010)] G. W. Hunt, A. Tordesillas, S. C. Green, and J. Shi, *Force-chain buckling in granular media: A structural mechanics perspective*, Philos. Trans. Royal Soc. A 368, 249–262 (2010).

[Hutzler *et al.* (1997)] S. Hutzler, D. Weaire, and R. Crawford, *Moving boundaries in ordered cylindrical foam structures*, Philos. Mag. B 75, 845–857 (1997).

[Hutzler *et al.* (1998)] S. Hutzler, D. Weaire, and R. Crawford, *Convective instability in foam drainage*, EPL 41, 461–466 (1998).

[Hutzler *et al.* (2014)] S. Hutzler, R. P. Murtagh, D. Whyte, S. T. Tobin, and D. Weaire, *Z-cone model for the energy of an ordered foam*, Soft Matter 10, 7103–7108 (2014).

[Hutzler *et al.* (2020)] S. Hutzler, A. Mughal, J. Ryan-Purcell, A. Irannezhad, and D. Weaire, *Buckling of a linear chain of hard spheres in a harmonic confining potential: Numerical and analytical results for low and high compression*, Phys. Rev. E 102, 022905 (2020).

[Jin *et al.* (2020)] W. Jin, H.-K. Chan, and Z. Zhong, *Shape-anisotropy-induced ordered packings in cylindrical confinement*, Phys. Rev. Lett. 124, 248002 (2020).

[Jin *et al.* (2021)] W. Jin, Y. Wang, H.-K. Chan and Z. Zhong, *Confinement-induced columnar crystals of ellipses*, Phys. Rev. Research 3, 013053 (2021).

[Johnson (1985)] K. L Johnson, *Contact Mechanics* (Cambridge University Press, Cambridge, 1985).

[Jones *et al.* (2001)] E. Jones, T. Oliphant, P. Peterson, et al., *SciPy: Open source scientific tools for Python* (2001). `http://www.scipy.org/`

[Jorjadze *et al.*(2011)] I. Jorjadze, L.-L. Pontani, K. A. Newhall, and J. Brujić, *Attractive emulsion droplets probe the phase diagram of jammed granular matter*, Proc. Natl. Acad. Sci. U.S.A. 108, 4286–4291 (2011).

[Katgert and van Hecke (2010)] G. Katgert and M. van Hecke, *Jamming and geometry of two-dimensional foams*, EPL 92, 34002 (2010).

[Khlobystov *et al.* (2004)] A. N. Khlobystov, D. A. Britz, A. Ardavan, and G. A. D. Briggs, *Observation of ordered phases of fullerenes in carbon nanotubes*, Phys. Rev. Lett. 92, 245507 (2004).

[Koch and Fischer (2004)] E. Koch and W. Fischer, *Sphere packings and packings of ellipsoids*, in *International Tables for Crystallography Volume C: Mathematical, Physical and Chemical tables* (Edited by E. Prince) (Springer, Netherlands, 2004), pp. 746–751.

[Koehler *et al.* (2000)] S. A. Koehler, S. Hilgenfeldt, and H. A. Stone, *A generalized view of foam drainage: Experiment and theory*, Langmuir 16, 6327–6341 (2000).

[Koga and Tanaka (2006)] K. Koga and H. Tanaka, *Close-packed structures and phase diagram of soft spheres in cylindrical pores*, J. Chem. Phys. 124, 131103 (2006).

[Lacasse *et al.* (1996)] M. D. Lacasse, G. S. Grest, and D. Levine, *Deformation of small compressed droplets*, Phys. Rev. E 54, 5436–5446 (1996).

[Landa *et al.* (2013)] H. Landa, R. Reznik, J. Brox, M. Mielenz, and T. Schätz, *Structure, dynamics and bifurcations of discrete solitons in trapped ion crystals*, New J. Phys. 15, 093003 (2013).

[Lázaro *et al.* (2018)] G. R. Lázaro, B. Dragnea, and M. F. Hagan, *Self-assembly of convex particles on spherocylindrical surfaces*, Soft Matter 14, 5728–5740 (2018).

[Lee *et al.* (2017)] T. Lee, K. Gizynski, and B. A. Grzybowski, *Non-equilibrium self-assembly of monocomponent and multicomponent tubular structures in rotating fluids*, Adv. Mater. 29, 1704274 (2017).

[Legoas *et al.* (2011)] S. B. Legoas, R. P. B. dos Santos, K. S. Troche, V. R. Coluci, and D. S. Galvão, *Ordered phases of encapsulated diamondoids into carbon nanotubes*, Nanotechnology 22, 315708 (2011).

[Liang *et al.* (2018)] Z. Liang, Y. Liu, F. Zhang, Y. Ai, and Q. Liang, *Dehydration-triggered shape morphing based on asymmetric bubble hydrogel microfibers*, Soft Matter 14, 6623–6626 (2018).

[Livio (2008)] M. Livio, *The golden ratio: The story of PHI, the world's most astonishing number* (Crown/Archetype, 2008).

[Lohr *et al.* (2010)] M. A. Lohr, A. M. Alsayed, B. G. Chen, Z. Zhang, R. D. Kamien, and A. G. Yodh, *Helical packings and phase transformations of soft spheres in cylinders*, Phys. Rev. E 81, 040401 (2010).

[Lubachevsky and Graham (2003)] B. D. Lubachevsky and R. Graham, *Dense packings of congruent circles in rectangles with a variable aspect ratio*, in *Discrete and Computational Geometry: The Goodman-Pollack Festschrift* (Edited by B. Aronov, S. Basu, J. Pach, and M. Sharir) (Springer, Berlin and Heidelberg, 2003), pp. 633–650.

[Ma (2020)] P. Ma, *A theoretical study of the electrical resistivities of densest-packed crystal structures of spheres in cylindrical confinement*, MSc thesis (2020).

[Ma and Chan (2021)] P. Ma and H.-K. Chan, *Densest-packed columnar structures of hard spheres: An investigation of the structural dependence of electrical conductivity*, Frontiers in Physics 9, 778001 (2021).

[Majmudar and Behringer (2005)] T. S. Majmudar and R. P. Behringer, *Contact force measurements and stress-induced anisotropy in granular materials*, Nature 435, 1079–1082 (2005).

[Manoharan (2019)] V. N. Manoharan, *Manoharan Lab - Optical metamaterials*, `https://manoharan.seas.harvard.edu/self-assembled-metamaterials` (Visited on 2019-07-23).

[Meagher *et al.* (2015)] A. J. Meagher, F. García-Moreno, J. Banhart, A. Mughal, and S. Hutzler, *An experimental study of columnar crystals using monodisperse microbubbles*, Colloids Surf. A 473, 55–59 (2015).

[Melzer (2006)] A. Melzer, *Zigzag transition of finite dust clusters*, Phys. Rev. E 73, 056404 (2006).

[Mermin and Wagner (1966)] N. D. Mermin and H. Wagner, *Absence of ferromagnetism or antiferromagnetism in one- or two-dimensional isotropic Heisenberg models*, Phys. Rev. Lett. 17, 1133–1136 (1966).

[Metropolis *et al.* (1953)] N. Metropolis, A. W. Rosenbluth, M. N. Rosenbluth, A. H. Teller, and E. Teller, *Equation of state calculations by fast computing machines*, J. Chem. Phys. 21, 1087–1092 (1953).

[Michaels *et al.* (2019)] T. C. T. Michaels, R. Kusters, A. J. Dear, C. Storm, J. C. Weaver, and L. Mahadevan, *Geometric localization in supported elastic struts*, Proc. R. Soc. A 475, 20190370 (2019).

[Mielenz *et al.* (2013)] M. Mielenz, J. Brox, S. Kahra, G. Leschhorn, M. Albert, T. Schaetz, H. Landa, and B. Reznik, *Trapping of topological-structural defects in Coulomb crystals*, Phys. Rev. Lett. 110, 133004 (2013).

[Morse and Witten (1993)] D. C. Morse and T. A. Witten, *Droplet elasticity in weakly compressed emulsions*, EPL 22, 549–555 (1993).

[Mughal (2013)] A. Mughal, *Screw symmetry in columnar crystals*, Philos. Mag. 93, 4070–4077 (2013).

[Mughal and Weaire (2014)] A. Mughal and D. Weaire, *Theory of cylindrical dense packings of disks*, Phys. Rev. E 89, 042307 (2014).

[Mughal *et al.* (2011)] A. Mughal, H.-K. Chan, and D. Weaire, *Phyllotactic description of hard sphere packing in cylindrical channels*, Phys. Rev. Lett. 106, 115704 (2011).

[Mughal *et al.* (2012)] A. Mughal, H.-K. Chan, D. Weaire, and S. Hutzler, *Dense packings of spheres in cylinders: Simulations*, Phys. Rev. E 85, 051305 (2012).

[Mughal *et al.* (2018)] A. Mughal, J. Winkelmann, D. Weaire, and S. Hutzler, *Columnar structures of soft spheres: Metastability and hysteresis*, Phys. Rev. E 98, 043303 (2018).

[Nigmatullin *et al.* (2016)] R. Nigmatullin, A. del Campo, G. De Chiara, G. Morigi, M. B. Plenio, and A. Retzker, *Formation of helical ion chains*, Phys. Rev. B 93, 014106 (2016).

[Nocedal and Wright (1999)] J. Nocedal and S. Wright, *Numerical Optimization* (Springer-Verlag, New York, 1999).

[Norman *et al.* (2018)] J. Norman, E. L. Sorrell, Y. Hu, V. Siripurapu, J. Garcia, J. Bagwell, P. Charbonneau, S. R. Lubkin, and M. Bagnat, *Tissue self-organization underlies morphogenesis of the notochord*, Philos. Trans. Royal Soc. B 373, 1759 (2018).

[Odriozola (2012)] G. Odriozola, *Revisiting the phase diagram of hard ellipsoids*, J. Chem. Phys. 136, 134505 (2012).

[Oğuz *et al.* (2009)] E. C. Oğuz, R. Messina, and H. Löwen, *Crystalline multilayers of the confined Yukawa system*, EPL 86, 28002 (2009).

[O'Hern *et al.* (2002)] C. S. O'Hern, S. A. Langer, A. J. Liu, and S. R. Nagel, *Random packings of frictionless particles*, Phys. Rev. Lett. 88, 075507 (2002).

[O'Hern *et al.* (2003)] C. S. O'Hern, L. E. Silbert, A. J. Liu, and S. R. Nagel, *Jamming at zero temperature and zero applied stress: The epitome of disorder*, Phys. Rev. E 68, 011306 (2003).

[O'Keeffe (1998)] M. O'Keeffe, *Sphere packings and space filling by congruent simple polyhedra*, Acta Crystallogr. A 54, 320–329 (1998).

[Ono-dit-Biot *et al.* (2020)] J.-C. Ono-dit-Biot, P. Soulard, S. Barkley, E. R. Weeks, T. Salez, E. Raphaël, and K. Dalnoki-Veress, *Rearrangement of two dimensional aggregates of droplets under compression: Signatures of the energy landscape from crystal to glass*, Phys. Rev. Research 2, 023070 (2020).

[Ozbay (2008)] E. Ozbay, *The magical world of photonic metamaterials*, Opt. Photon. News 19, 22–27 (2008).

[Partner *et al.* (2015)] H. L. Partner, R. Nigmatullin, T. Burgermeister, J. Keller, K. Pyka, M. B. Plenio, A. Retzker, W. H. Zurek, A. del Campo, and T. E. Mehlstäubler, *Structural phase transitions and topological defects in ion Coulomb crystals*, Physica B: Condensed Matter 460, 114–118 (2015).

[Pickett *et al.* (2000)] G. T. Pickett, M. Gross, and H. Okuyama, *Spontaneous chirality in simple systems*, Phys. Rev. Lett. 85, 3652–3655 (2000).

[Pittet *et al.* (1995)] N. Pittet, N. Rivier, and D. Weaire, *Cylindral packing of foam cells*, Forma 10, 65–73 (1995).

[Pittet *et al.* (1996)] N. Pittet, P. Boltenhagen, N. Rivier, and D. Weaire, *Structural transitions in ordered, cylindrical foams*, EPL 35, 547–552 (1996).

[Prusinkiewicz and Lindenmeyer (1990)] P. Prusinkiewicz and A. Lindenmeyer, *Phyllotaxis* (book chapter), in *The Algorithmic Beauty of Plants* (Springer-Verlag, 1990), pp. 99–118. `http://algorithmicbotany.org/papers/#abop`

[Pyka *et al.* (2013)] K. Pyka, J. Keller, H. L. Partner, R. Nigmatullin, T. Burgermeister, D. M. Meier, K. Kuhlmann, A. Retzker, M. B. Plenio, W. H. Zurek, A. del Campo, and T. E. Mehlstäubler, *Topological defect formation and spontaneous symmetry breaking in ion Coulomb crystals*, Nat. Commun. 4, 2291 (2013).

[Redhill Precision Speciality Balls (Manufacturer)] Redhill Precision Speciality Balls, `https://www.redhill-balls.eu` (Visited on 2019-07-23).

[Rogava (2019)] A. Rogava, *Tennis-ball towers*, Physics World 32, 25 (2019).

[Rogers (1964)] C. A. Rogers, *Packing and Covering* (Cambridge University Press, Cambridge, 1964).

[Saadatfar *et al.* (2008)] M. Saadatfar, J. D. Barry, D. Weaire, and S. Hutzler, *Ordered cylindrical foam structures with internal bubbles*, Philos. Mag. 88, 661–668 (2008).

[Saba *et al.* (2013)] M. Saba, M. D. Turner, K. Mecke, M. Gu, and G. E. Schröder-Turk, *Group theory of circular-polarization effects in chiral photonic crystals with four-fold rotation axes applied to the eight-fold intergrowth of gyroid nets*, Phys. Rev. B 88, 245116 (2013).

[Sadoc *et al.* (2012)] J.-F. Sadoc, N. Rivier, and J. Charvolin, *Phyllotaxis: A non-conventional crystalline solution to packing efficiency in situations with radial symmetry*, Acta Crystallogr. A 68, 470–483 (2012).

[Sanwaria *et al.* (2014)] S. Sanwaria, A. Horechyy, D. Wolf, C.-Y. Chu, H.-L. Chen, P. Formanek, M. Stamm, R. Srivastava, and B. Nandan, *Helical packing of nanoparticles confined in cylindrical domains of a self-assembled block copolymer structure*, Angew. Chem. Int. Ed. 53, 9090–9093 (2014).

[Schade *et al.* (2013)] N. B. Schade, M. C. Holmes-Cerfon, E. R. Chen, D. Aronzon, J. W. Collins, J. A. Fan, F. Capasso, and V. N. Manoharan, *Tetrahedral colloidal clusters from random parking of bidisperse spheres*, Phys. Rev. Lett. 110, 148303 (2013).

[Schroeder-Turk *et al.* (2015)] G. Schroeder-Turk, J. Knauf, R. Roth, and K. Mecke, *Can a hard-sphere fluid feel the topology of a confining pore?*, Bull. Am. Phys. Soc. 60 (2015).

[Shalaev (2007)] V. M. Shalaev, *Optical negative-index metamaterials*, Nat. Photonics 1, 41–48 (2007).

[Smalley (2003)] R. E. Smalley, *Forward*, in *Carbon Nanotubes: Synthesis, Structure, Properties, and Applications* (Edited by M. S. Dresselhaus, G. Dresselhaus, and P. Avouris) (Springer, 2001).

[Srebnik and Douglas (2011)] S. Srebnik and J. F. Douglas, *Self-assembly of charged particles on nanotubes and the emergence of particle rings, chains, ribbons and chiral sheets*, Soft Matter 7, 6897–6905 (2011).

[Stoyan and Yaskov (2010)] Y. G. Stoyan and G. N. Yaskov, *Packing identical spheres into a cylinder*, Int. T. Oper. Res. 17, 51–70 (2010).

[Straube *et al.* (2011)] A. V. Straube, A. A. Louis, J. Baumgartl, C. Bechinger, and R. P. A. Dullens, *Pattern formation in colloidal explosions*, EPL 94, 48008 (2011).

[Straube *et al.* (2013)] A. V. Straube, R. P. A. Dullens, L. Schimansky-Geier, and A. A. Louis, *Zigzag transitions and nonequilibrium pattern formation in colloidal chains*, J. Chem. Phys. 139, 134908 (2013).

[Takezoe and Takanishi (2006)] H. Takezoe and Y. Takanishi, *Bent-core liquid crystals: Their mysterious and attractive world*, Jpn. J. Appl. Phys. 45, 597–625 (2006).

[Tanjeem (2020)] N. Tanjeem, *Effect of geometry on packing and parking of colloidal spheres*, PhD thesis (2020).

[Thompson (1917)] D. W. Thompson, *On Growth and Form*, 1st Ed. (Cambridge University Press, Cambridge, 1917).

[Thompson (1983)] T. M. Thompson, *From Error-Correcting Codes Through Sphere Packings to Simple Groups* (Mathematical Association of America, New York, 1983).

[Thompson (2015)] R. C. Thompson, *Ion coulomb crystals*, Contemp. Phys. 56, 63–79 (2015).

[Tobin *et al.* (2011)] S. T. Tobin, J. D. Barry, A. J. Meagher, B. Bulfin, C. E. O'Rathaille, and S. Hutzler, *Ordered polyhedral foams in tubes with circular, triangular and square cross-section*, Colloids Surf. A 382, 24–31 (2011).

[Torquato and Jiao (2009)] S. Torquato and Y. Jiao, *Dense packings of the Platonic and Archimedean solids*, Nature 460, 876–879 (2009).

[Torquato and Stillinger (2010)] S. Torquato and F. H. Stillinger, *Jammed hard-particle packings: From Kepler to Bernal and beyond*, Rev. Mod. Phys. 82, 2633 (2010).

[Troche *et al.* (2005)] K. S. Troche, V. R. Coluci, S. F. Braga, D. D. Chinellato, F. Sato, S. B. Legoas, R. Rurali, and D. S. Galvão, *Prediction of ordered phases of encapsulated C60, C70, and C78 inside carbon nanotubes*, Nano Lett. 5, 349–355 (2005).

[Tsytovich *et al.* (2007)] V. N. Tsytovich, G. E. Morfill, V. E. Fortov, N. G. Gusein-Zade, B. A. Klumov, and S. V. Vladimirov, *From plasma crystals and helical structures towards inorganic living matter*, New J. Phys. 9, 263 (2007).

[Turner *et al.* (2013a)] M. D. Turner, M. Saba, Q. Zhang, B. P. Cumming, G. E. Schröder-Turk, and M. Gu, *Miniature chiral beamsplitter based on gyroid photonic crystals*, Nat. Photonics 7, 801–805 (2013).

[Turner *et al.* (2013b)] M. D. Turner, M. Saba, Q. Zhang, B. P. Cumming, G. E. Schröder-Turk, and M. Gu, *A chiral beamsplitter inspired by butterfly nanostructures*, Frontiers in Optics 2013 Postdeadline, FW6A.3 (2013).

[Tymczenko *et al.* (2008)] M. Tymczenko, L. F. Marsal, T. Trifonov, I. Rodriguez, F. Ramiro-Manzano, J. Pallares, A. Rodriguez, R. Alcubilla, and F. Meseguer, *Colloidal crystal wires*, Adv. Mater 20, 2315–2318 (2008).

[van Hecke (2010)] M. van Hecke, *Jamming of soft particles: Geometry, mechanics, scaling and isostaticity*, J. Phys.: Condens. Matter 22, 033101 (2010).

[Wales and Doye (1997)] D. J. Wales and J. P. K. Doye, *Global optimization by basin-hopping and the lowest energy structures of Lennard-Jones clusters containing up to 110 Atoms*, J. Phys. Chem. A 101, 5111–5116 (1997).

[Wang *et al.* (2016)] K. Wang, S.-M. Jin, J. Xu, R. Liang, K. Shezad, Z. Xue, X. Xie, E. Lee, and J. Zhu, *Electric-field-assisted assembly of polymer-*

tethered gold nanorods in cylindrical nanopores, ACS Nano 10, 4954–4960 (2016).

[Warner and Wilson (2010)] J. H. Warner and M. Wilson, *Elastic distortions of carbon nanotubes induced by chiral fullerene chains*, ACS Nano 4, 4011–4016 (2010).

[Weaire (1996)] D. Weaire, *The Kelvin problem: Foam Structures of Minimal Surface Area* (Taylor and Francis, London, 1996).

[Weaire and Phelan (1994)] D. Weaire and R. Phelan, *A counterexample to Kelvin's conjecture on minimal surfaces*, Phil. Mag. Lett. 69, 107–110 (1994).

[Weaire *et al.* (1993)] D. Weaire, N. Pittet, S. Hutzler, and D. Pardal, *Steady-state drainage of an aqueous foam*, Phys. Rev. Lett. 71, 2670–2673 (1993).

[Weaire *et al.* (1997)] D. Weaire, S. Hutzler, G. Verbist, and E. Peters, *A review of foam drainage*, Adv. Chem. Phys. 102, 315–374 (1997).

[Weaire *et al.* (2017)] D. Weaire, R. Höhler, and S. Hutzler, *Bubble-bubble interactions in a 2d foam, close to the wet limit*, Adv. Colloid Interface Sci. 247, 491–495 (2017).

[Weaire *et al.* (2020)] D. Weaire, A. Irannezhad, A. Mughal, and S. Hutzler, *A simple experimental system to illustrate the nonlinear properties of a linear chain under compression*, Am. J. Phys. 88, 347–352 (2020).

[Winkelmann (2019)] J. Winkelmann, *Ordered columnar structures: Golf balls packed inside tube*, `https://www.reddit.com/r/ScienceGIFs/comments/b2c604/ordered_columnar_structures_golf_balls_packed/` (Visited on 2019-07-23).

[Winkelmann (2020)] J. Winkelmann, *Structures of columnar packings with soft and hard spheres*, PhD thesis (2020).

[Winkelmann and Chan (2020)] J. Winkelmann and H.-K. Chan, *Repository for simulations of columnar structures.* `https://bitbucket.org/jWinman/sequential-deposition-code/src/master/`

[Winkelmann (2021)] J. Winkelmann, *Sphere packing in a cylinder*, `https://en.wikipedia.org/wiki/Cylinder_sphere_packing` (Visited on 2021-06-24).

[Winkelmann *et al.* (2017)] J. Winkelmann, F. F. Dunne, V. J. Langlois, M. E. Möbius, D. Weaire, and S. Hutzler, *2D foams above the jamming transition: Deformation matters*, Colloids Surf. A 534, 52–57 (2017).

[Winkelmann *et al.* (2018)] J. Winkelmann, B. Haffner, D. Weaire, A. Mughal, and S. Hutzler, *Corrected article: Simulation and observation of line-slip structures in columnar structures of soft spheres [Phys. Rev. E 96, 012610 (2017)]*, Phys. Rev. E 97 059902 (2018).

[Winkelmann *et al.* (2019a)] J. Winkelmann, A. Mughal, D. B. Williams, D. Weaire, and S. Hutzler, *Theory of rotational columnar structures of soft spheres*, Phys. Rev. E 99, 020602(R) (2019).

[Winkelmann *et al.* (2019b)] J. Winkelmann, A. Mughal, D. Weaire, and S. Hutzler, *Equilibrium configurations of hard spheres in a cylindrical harmonic potential*, EPL 127, 44002 (2019).

[Wood (2017)] D. Wood, *Assembly of particles onto rigid cylinders and flexible membranes: Probing effects of surface curvature and deformation*, PhD thesis (2017).

[Wood *et al.* (2013)] D. A. Wood, C. D. Santangelo, and A. D. Dinsmore, *Self-assembly on a cylinder: A model system for understanding the constraint of commensurability*, Soft Matter 9, 10016–10024 (2013).

[Wu *et al.* (2004)] Y. Wu, G. Cheng, K. Katsov, S. W. Sides, J. Wang, J. Tang, G. H. Fredrickson, M. Moskovits, and G. D. Stucky, *Composite mesostructures by nano-confinement*, Nat. Mat. 3, 816 (2004).

[Wu *et al.* (2017)] G. Wu, H. Cho, D. A. Wood, A. D. Dinsmore, and S. Yang, *Confined assemblies of colloidal particles with soft repulsive interactions*, J. Am. Chem. Soc. 139, 5095–5101 (2017).

[www.cambridge2000.com] `http://www.cambridge2000.com/gallery/html/PA309665e.html` (Visited on 2019-08-12).

[Yamazaki *et al.* (2008)] T. Yamazaki, K. Kuramochi, D. Takagi, Y. Homma, F. Nishimura, N. Hori, K. Watanabe, S. Suzuki, and Y. Kobayashi, *Ordered fullerene nanocylinders in large-diameter carbon nanotubes*, Nanotechnology 19, 045702 (2008).

[Yamchi and Bowles (2015)] M. Z. Yamchi and R. K. Bowles, *Helical defect packings in a quasi-one-dimensional system of cylindrically confined hard spheres*, Phys. Rev. Lett. 115, 025702 (2015).

[Yan *et al.* (2016)] L. L. Yan, W. Wan, L. Chen, F. Zhou, S. J. Gong, X. Tong, and M. Feng, *Exploring structural phase transitions of ion crystals*, Sci. Rep. 6, 21547 (2016).

[Yin and Xia (2003)] Y. Yin and Y. Xia, *Self-assembly of spherical colloids into helical chains with well-controlled handedness*, J. Am. Chem. Soc. 125, 2048–2049 (2003).

[Zerrouki *et al.* (2008)] D. Zerrouki, J. Baudry, D. Pine, P. Chaikin, and J. Bibette, *Chiral colloidal clusters*, Nature 455, 380–382 (2008).

[Zhang *et al.* (2019)] Q. Zhang, J. Gu, L. Zhang, and J. Lin, *Diverse chiral assemblies of nanoparticles directed by achiral block copolymers via nanochannel confinement*, Nanoscale 11, 474–484 (2019).

Index